SCHAUM'S *Easy* OUTLINES

COLLEGE
MATHEMATICS

Other Books in Schaum's Easy Outlines Series Include:

SCHAUM'S *Easy* OUTLINES

COLLEGE
MATHEMATICS

BASED ON SCHAUM'S
Theory and Problems of
College Mathematics
BY FRANK AYRES, JR.
AND PHILIP A. SCHMIDT

ABRIDGEMENT EDITOR:
GEORGE J. HADEMENOS

SCHAUM'S OUTLINE SERIES
McGRAW-HILL

New York Chicago San Francisco Lisbon London Madrid
Mexico City Milan Montreal New Delhi San Juan
Seoul Singapore Sydney Toronto

FRANK AYRES, JR. was formerly Professor and Head of the Department of Mathematics at Dickinson College, Carlisle, Pennsylvania.

PHILIP A. SCHMIDT has a B.S. from Brooklyn College and both M.A. and Ph.D. degrees from Syracuse University. He is Associate Provost for Academic Services at Berea College in Kentucky and previously Dean of the School of Education at SUNY at New Paltz.

GEORGE J. HADEMENOS has taught at the University of Dallas and done research at the University of Massachusetts Medical Center and the University of California at Los Angeles. He holds a B.S. from Angelo State University and both M.S. and Ph.D. degrees from the University of Texas at Dallas. He is the author of several books in the Schaum's Outline series.

3 4 5 6 7 8 9 10 11 12 13 14 15 DOC DOC 0 9 8 7 6 5 4

Library of Congress Cataloging-in-Publication Data

Ayres, Frank, 1901-
    College mathematics : based on Schaum's Theory and problems of college mathematics / by Frank Ayres, Jr. and Philip A. Schmidt ; abridgement editor, George J. Hademenos.
        p. cm. — (Schaum's outline series) (Schaum's easy outlines)
    Includes index.
    ISBN 0-07-136975-9 (pbk.)
    1. Mathematics—Outlines, syllabi, etc. I. Schmidt, Philip A. II. Ayres, Frank, 1901- Schaum's outline theory and problems of college mathematics. III. Hademenos, George J. IV. Title. V. Series.

QA37.3 .A97 2001
510—dc21   2001030767

McGraw-Hill

*A Division of The **McGraw-Hill** Companies*

Contents

Chapter 1
FUNDAMENTALS OF ALGEBRA

IN THIS CHAPTER:

✔ *The Number System of Algebra*
✔ *Elements of Algebra*
✔ *Inequalities*
✔ *Logarithms*
✔ *Power, Exponential, and Logarithmic Curves*

The Number System of Algebra

Elementary mathematics is concerned mainly with certain elements called *numbers* and with certain operations defined on them.

The unending set of symbols 1, 2, 3, 4, 5, 6, 7, 8, 9, 10, 11, 12, … used in counting are called *natural numbers*.

In adding two of these numbers, say 5 and 7, we begin with 5 (or with 7) and count to the *right* seven (or five) numbers to get 12. The sum of two natural numbers is a natural number, that is, the sum of two members of the above set is a member of the set.

In subtracting 5 from 7, we begin with 7 and count to the *left* five numbers to 2. It is clear, however, that 7 cannot be subtracted from 5 using only natural numbers since there are only four numbers to the left of 5.

Integers

In order that subtraction be always possible, it is necessary to increase our set of numbers. We prefix each natural number with a + sign (in practice, it is more convenient not to write the sign) to form the *positive integers*, we prefix each natural number with a − sign (the sign must always be written) to form the *negative integers*, and we create a new symbol 0, read *zero* as shown in Fig. 1-1.

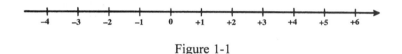

Figure 1-1

On the set of *integers*

$$..., -8, -7, -6, -5, -4, -3, -2, -1, 0, +1, +2, +3, +4, +5, +6, +7, +8,...$$

the operations of addition and subtraction are possible without exception.

Rule 1. To add two numbers having like signs, add their numerical values and prefix their common sign.

Rule 2. To add two numbers having unlike signs, subtract the smaller numerical value from the larger, and prefix the sign of the number having the larger numerical value.

Rule 3. To subtract a number, change its sign and add.

Rule 4. To multiply or divide two numbers (never divide by 0!), multiply or divide the numerical values, prefixing a + sign if the two numbers have like signs and a − sign if the two numbers have unlike signs.

Rational Numbers

The set of rational numbers consists of all numbers of the form m/n, where m and $n \neq 0$ are integers. Thus, the rational numbers include the integers and common fractions, as shown in Fig. 1-2. A fraction is said to be expressed in lowest terms by the representation m/n where m and n have no common prime factor. Other rules concerning rational numbers are, therefore

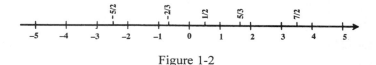

Figure 1-2

Rule 5. The value of a rational number is unchanged if both the numerator and denominator are multiplied or divided by the same nonzero number.

Rule 6. The sum (difference) of two rational numbers expressed with the same denominator is a rational number whose denominator is the common denominator and whose numerator is the sum (difference) of the numerators.

Rule 7. The product of two or more rational numbers is a rational number whose numerator is the product of the numerators and whose denominator is the product of the several factors.

Rule 8. The quotient of two rational numbers can be evaluated by the use of Rule 5 with the least common denominator of the two numbers as the multiplier.

If a and b are rational numbers, $a + b$, $a - b$, and $a \cdot b$ are rational numbers. Moreover, if a and b are $\neq 0$, there exists a rational number x, unique except for its representation, such that

$$ax = b$$

When a or b or both are zero, we have the following situations:

If $b = 0$ and $a \neq 0$ in $ax = b$, then $a \cdot x = 0$ implying that $x = 0$; that is, $0/a = 0$ when $a \neq 0$.

If $a = 0$ and $b \neq 0$ in $ax = b$, then $0 \cdot x = b$; and, $b/0$, when $b \neq 0$, is without meaning since $0 \cdot x = 0$.

If $a = 0$ and $b = 0$ in $ax = b$, then $0 \cdot x = 0$; and, $0/0$ is indeterminate since every number x satisfies the equation.

In brief: $0/a = 0$ when $a \neq 0$, but division by 0 is never permitted.

Decimals

In writing numbers, we use a positional system, that is, the value given any particular digit depends upon its position in the sequence. For example, in 423, the positional value of the digit 4 is 4(100) while in 234, the positional value of the digit 4 is 4(1). Since the positional value of a digit involves the number 10, this system of notation is called the *decimal system*. In this system, the number 4238.75 means

$$4(1000) + 2(100) + 3(10) + 8(1) + 7\left(\frac{1}{10}\right) + 5\left(\frac{1}{100}\right)$$

It is interesting to note that, from this example, certain definitions to be made later involving exponents may be anticipated. Since $1000 = 10^3$, $100 = 10^2$, $10 = 10^1$, it would seem natural to define $1 = 10^0$,

$$\frac{1}{10} = 10^{-1}, \text{ and } \frac{1}{100} = 10^{-2}.$$

Percentage

The symbol %, read percent, means per hundred; thus 5% is equivalent to $\dfrac{5}{100}$ or 0.05. Any number, when expressed in decimal notation, can be written as a percent by multiplying by 100 and adding the symbol %. Conversely, any percentage may be expressed in decimal form by dropping the symbol % and dividing by 100.

Irrational Numbers

The existence of numbers other than the rational numbers may be inferred from either of the following considerations:

(a) We may conceive of a nonrepeating decimal constructed in endless time by setting down a succession of digits chosen at random.
(b) The lengthss of the diagonal of a square of side 1 is not a rational number, that is, there exists, no rational number a such that $a^2 = 2$. Numbers such as $\sqrt{2}$, $\sqrt[3]{2}$, $\sqrt[5]{-3}$, and π (but not $\sqrt{-3}$) are called *irrational numbers*. The first three of these are called *radicals*. The radical $\sqrt[n]{a}$ is said to be of order n; n is called the *index* and a is called the *radicand*.

Real Numbers

The set of *real numbers* consists of the rational and irrational numbers, as shown in Fig. 1-3. The real numbers may be ordered by comparing their decimal representations.

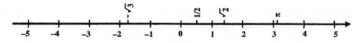

Figure 1-3

The number associated with a point on the line, called the *coordinate* of the point, gives its distance and direction from that point (called the origin) associated with the number 0. If a point A has coordinate a, we shall speak of it as the point A (a).

Complex Numbers

In the set of real numbers, there is no number whose square is -1. If there is to be such a number, say $\sqrt{-1}$, then by definition $\left(\sqrt{-1}\right)^2 = -1$. Note carefully that

$$\left(\sqrt{-1}\right)^2 = \sqrt{-1}\sqrt{-1} = \sqrt{(-1)(-1)} = \sqrt{1} = 1$$

is incorrect. In order to avoid this error, the symbol i with the following properties is used:

If $a > 0$, $\qquad \sqrt{-a} = i\sqrt{a}$; $\qquad\qquad i^2 = -1$

Then

$$\left(\sqrt{-2}\right)^2 = \sqrt{-2}\sqrt{-2} = \left(i\sqrt{2}\right)\left(i\sqrt{2}\right) = i^2 \cdot 2 = -2$$

and

$$\sqrt{-2}\sqrt{-3} = \left(i\sqrt{2}\right)\left(i\sqrt{3}\right) = i^2\sqrt{6} = -\sqrt{6}$$

Numbers of the form $a + bi$, where a and b are real numbers, are called *complex numbers*. In the complex number $a + bi$, a is called the *real part* and bi is called the *imaginary part*. Numbers of the form ci, where c is real, are called *imaginary numbers*.

The complex number $a + bi$ is a real number when $b = 0$ and a pure imaginary number when $a = 0$. When a complex number is not a real number, it is called *imaginary*.

To add (subtract) two complex numbers, add (subtract) the real parts and add (subtract) the pure imaginary parts.

To multiply two complex numbers, form the product treating i as an ordinary number and then replace i^2 by -1.

Elements of Algebra

Positive Integral Exponents

If a is any number and n is any positive integer, the product of the n factors $a \cdot a \cdot a \cdots a$ is denoted by a^n. To distinguish between the letters, a is called the *base* and n is called the *exponent*.

If a and b are any bases and m and n are any positive integers, we have the following laws of exponents:

(1) $\quad a^m \cdot a^n = a^{m+n}$

(2) $\quad \left(a^m\right)^n = a^{mn}$

(3) $\quad \dfrac{a^m}{a^n} = a^{m-n}, \; a \neq 0, \; m > n; \quad \dfrac{a^m}{a^n} = \dfrac{1}{a^{n-m}}, \; a \neq 0, \; m < n$

(4) $\quad \left(a \cdot b\right)^n = a^n b^n$

(5) $\quad \left(\dfrac{a}{b}\right)^n = \dfrac{a^n}{b^n}, \; b \neq 0$

Let n be a positive integer and a and b be two numbers such that $b^n = a$; then b is called an nth *root of a*. Every number $a \neq 0$ has exactly n distinct nth roots.

If a is imaginary, all of its nth roots are imaginary.

If a is real and n is odd, then exactly one of the nth roots of a is real.

If a is real and n is even, then there are exactly two real nth roots of a when $a > 0$, but no real nth roots of a when $a < 0$.

The principal nth root of a is the positive real nth root when a is positive and the real nth root of a, if any, when a is negative. The principal nth root of a is denoted by $\sqrt[n]{a}$, called a *radical*. The integer n is called the *index* of the radical and a is called the *radicand*.

Zero, Fractional, and Negative Exponents

When s is a positive integer, r is any integer, and p is any rational number, the following extend the definition of a^n in such a way that the laws (1) – (5) are satisfied when n is any rational number.

(6) $a^0 = 1, \ a \neq 0$

(7) $a^{r/s} = \sqrt[s]{a^r} = \left(\sqrt[s]{a}\right)^r$

(8) $a^{-p} = \dfrac{1}{a^p}, \ a \neq 0$

Solved Problem 1-1. Evaluate: (a) $81^{1/2}$; (b) $81^{3/4}$; (c) $\left(\dfrac{16}{49}\right)^{3/2}$.

Solution

(a) $81^{1/2} = \sqrt{81} = 9$

(b) $81^{3/4} = \left(\sqrt[4]{81}\right)^3 = 3^3 = 27$

(c) $\left(\dfrac{16}{49}\right)^{3/2} = \left(\sqrt{\dfrac{16}{49}}\right)^3 = \left(\dfrac{4}{7}\right)^3 = \dfrac{64}{343}$

Solved Problem 1-2. Perform each of the following operations and express the result without negative or zero exponents.

Solution

(a) $\left(\dfrac{81a^4}{b^8}\right)^{-1/4} = \dfrac{3^{-1}a^{-1}}{b^{-2}} = \dfrac{b^2}{3a}$

(b) $\left(a^{1/2} + a^{-1/2}\right)^2 = a + 2a^0 + a^{-1} = a + 2 + \dfrac{1}{a}$

Inequalities

An *inequality* is a statement that one (real) number is greater or less than another; for example, $3 > -2$, $-10 < -5$.

Two inequalities are said to have the *same sense* if their signs of inequality point in the same direction. Thus, $3 > -2$ and $-5 > -10$ have the same sense; $3 > -2$ and $-10 < -5$ have opposite senses.

The sense of an equality is *not* changed:

(a) if the same number is added to or subtracted from both sides
(b) if both sides are multiplied or divided by the same *positive* number

The sense of an equality *is* changed if both sides are multiplied or divided by the same negative number.

An *absolute inequality* is one which is true for all real values of the letters involved; for example, $x^2 + 1 > 0$ is an absolute inequality.

A *conditional inequality* is one which is true for certain values of the letters involved; for example, $x + 2 > 5$ is a conditional inequality since it is true for $x = 4$ but not for $x = 1$.

The solution of a conditional inequality in one letter, say x, consists of all values of x for which the inequality is true. These values lie on one or more intervals of the real number scale.

To solve a linear inequality, proceed as in solving a linear equality keeping in mind the rules for keeping or reversing the sense.

Solved Problem 1-3. Solve $2x - 8 < 7x + 12$.

Solution. Subtract $7x - 8$ from each member:

$$-5x < 20$$

Divide by -5 :

$$x > -4$$

Graphical representation

See Fig. 1-4.

Figure 1-4

$$f(x) = ax^2 + bx + c > 0$$

To solve a quadratic inequality, $f(x)$, solve the equality $= 0$, locate the roots r_1 and r_2 on a number scale, and determine the sign of $f(x)$ on each of the resulting intervals.

Solved Problem 1-4. Solve $(x+5)(x-1)(x-2) < 0$.

Solution. Solve the equality

$$f(x) = (x+5)(x-1)(x-2) = 0. \quad x = 1, 2, -5$$

Locate the roots on a number scale.

Determine the sign of $f(x)$:

On the interval $x < -5$: $f(-6) = (-1)(-7)(-8) < 0$

On the interval $-5 < x < 1$: $f(0) = 5(-1)(-2) > 0$

On the interval $1 < x < 2$: $f\left(\dfrac{3}{2}\right) = \left(\dfrac{13}{2}\right)\left(\dfrac{1}{2}\right)\left(-\dfrac{1}{2}\right) < 0$

On the interval $x > 2$: $\qquad f(3) = 8 \cdot 2 \cdot 1 > 0$

The inequality is satisfied when $x < -5$ and $1 < x < 2$. See Fig. 1-5.

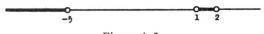

Figure 1-5

Logarithms

The *logarithm* of a positive number N to a given base b (written $\log_b N$) is the exponent of the power to which b must be raised to produce N. It is understood that b is positive and different from 1.

For example,

(a) Since $9 = 3^2$, $\log_3 9 = 2$.

(b) Since $64 = 4^3$, $\log_4 64 = 3$.

(c) Since $64 = 2^6$, $\log_2 64 = 6$.

(d) Since $1000 = 10^3$, $\log_{10} 1000 = 3$.

(e) Since $0.01 = 10^{-2}$, $\log_{10} 0.01 = -2$.

Note!

Fundamental Laws of Logarithms

1. The logarithm of the product of two or more positive numbers is equal to the sum of the logarithms of the several numbers. For example,

$$\log_b (P \cdot Q \cdot R) = \log_b P + \log_b Q + \log_b R$$

2. The logarithm of the quotient of two positive numbers is equal to the logarithm of the dividend minus the logarithm of the divisor. For example,

$$\log_b \frac{P}{Q} = \log_b P - \log_b Q$$

3. The logarithm of a power of a positive number is equal to the logarithm of the number, multiplied by the exponent of the power. For example,

$$\log_b P^n = n \log_b P$$

4. The logarithm of a root of a positive number is equal to the logarithm of the number, divided by the index of the root. For example,

$$\log_b \sqrt[n]{P} = \frac{1}{n} \log_b P$$

Solved Problem 1-5. Prove the four laws of logarithms.

Solution. Let $P = b^p$ and $Q = b^q$; then $\log_b P = p$ and $\log_b Q = q$.

1. Since $P \cdot Q = b^p \cdot b^q = b^{p+q}$, $\log_b PQ = p + q = \log_b P + \log_b Q$;

    that is, the logarithm of the product of two positive numbers is equal to the sum of the logarithms of the numbers.

2. Since $\dfrac{P}{Q} = \dfrac{b^p}{b^q} = b^{p-q}$, $\log_b \left(\dfrac{P}{Q}\right) = p - q = \log_b P - \log_b Q$;

    that is, the logarithm of the quotient of two positive numbers is the logarithm of the numerator minus the logarithm of the denominator.

3. Since $P^n = \left(b^p\right)^n = b^{np}$, $\log_b P^n = np = n\log_b P$; that is, the loga-
    rithm of a power of a positive number is equal to the product of the
    exponent and the logarithm of the number.

4. Since $\sqrt[n]{P} = P^{1/n} = b^{p/n}$, $\log_b \sqrt[n]{P} = \dfrac{p}{n} = \dfrac{1}{n}\log_b P$; that is, the log-
    arithm of a root of a positive number is equal to the logarithm of the
    number divided by the index of the root.

An exponential equation is an equation involving one or more
unknowns in an exponent. For example, $2^x = 7$ and $(1.03)^{-x} = 2.5$
are exponential equations. Such equations are solved by means of logarithms.

Solved Problem 1-6. Solve the exponential equation $2^x = 7$.

Solution. Take logarithms of both sides:

$$x\log 2 = \log 7$$

Solve for x:

$$x = \frac{\log 7}{\log 2} = \frac{0.8451}{0.3010}$$

Evaluate, using logarithms:

$$
\begin{aligned}
\log 0.8451 &= 9.9270 - 10 \\
- \log 0.3010 &= \underline{9.4786 - 10} \\
\log x \quad\;\; &= 0.4484
\end{aligned}
$$

$$x = 2.808$$

A useful system of logarithms is the *natural system* in which the base is a certain irrational number $e = 2.71828$, approximately.

The natural logarithm of N, ln N, and the common logarithm of N, log N, are related by the formula

ln $N = 2.03026$ log N

Solved Problem 1-7. Show that $b^{3\log_b x} = x^3$.

Solution. Let $3 \log_b x = t$. Then, $\log_b x^3 = t$ and $x^3 = b^t = b^{3\log_b x}$.

Power, Exponential, and Logarithmic Curves

Power functions in x are of the form x^n. If $n > 0$, the graph of $y = x^n$ is said to be of the *parabolic* type (the curve is a parabola for $n = 2$). If $n < 0$, the graph of $y = x^n$ is said to be of the *hyperbolic* type (the curve is a hyperbola for $n = -1$).

Solved Problem 1-8. Sketch the graphs of (a) $y = x^{3/2}$ and (b) $y = -x^{-3/2}$.

Solution. Table 1-1 has been computed for selected values of x. We shall assume that the points corresponding to intermediate values of x lie on a smooth curve joining the points given in the table. See Figs. 1-6 and 1-7.

x	$y = x^{3/2}$	$y = -x^{-3/2}$
9	27	$-\frac{1}{27}$
4	8	$-\frac{1}{8}$
1	1	-1
$\frac{1}{4}$	$\frac{1}{8}$	-8
$\frac{1}{9}$	$\frac{1}{27}$	-27
0	0	—

Table 1-1

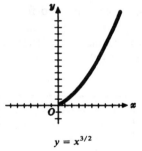

$$y = x^{3/2}$$

Figure 1-6

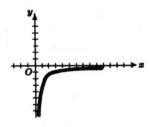

$$y = -x^{-3/2}$$

Figure 1-7

Exponential functions in x are of the form b^x where b is a constant. The discussion will be limited here to the case $b > 1$.

The curve whose equation is $y = b^x$ is called an *exponential curve*. The general properties of such curves are:

(a) The curve passes through the point $(0,1)$.
(b) The curve lies above the x axis and has that axis as an asymptote.

Solved Problem 1-9. Sketch the graphs of (a) $y = 2^x$ and (b) $y = 3^x$.

Solution. Table 1-2 has been computed for selected values of x. The exponential equation appears frequently in the form $y = ce^{kx}$ where c and k are nonzero constants and $e = 2.71828\ldots$ is the natural logarithmic base. See Figs. 1-8 and 1-9.

x	$y = 2^x$	$y = 3^x$
3	8	27
2	4	9
1	2	3
0	1	1
-1	$\frac{1}{2}$	$\frac{1}{3}$
-2	$\frac{1}{4}$	$\frac{1}{9}$
-3	$\frac{1}{8}$	$\frac{1}{27}$

Table 1-2

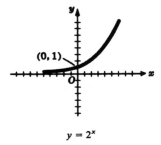

$y = 2^x$

Figure 1-8

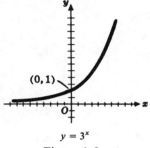

$$y = 3^x$$

Figure 1-9

The curve whose equation is $y = \log_b x, \; b > 1,$ is called a *logarithmic curve*. The general properties are:

(a) The curve passes through the point (1,0).
(b) The curve lies to the right of the y axis and has that axis as an asymptote.

Chapter 2
FUNCTIONS
AND EQUATIONS

Functions

A correspondence (x, y) between two sets of numbers which pairs to an arbitrary number y of the first set exactly one number x of the second set is called a *function*. In this case, it is customary to speak of y as a *function of* x. The variable x is called the *independent variable* and y is called the *dependent variable*.

18

A function may be defined:

(a) by a table of correspondents or table of values, as in Table 2-1:

x	1	2	3	4	5	6	7	8	9	10
y	3	4	5	6	7	8	9	10	11	12

Table 2-1

(b) by an equation or formula, as in $y = x + 2$. For each value assigned to x, the above relation yields a correspondent. Note that the table above is a table of values for this function.

A function is called *single-valued* if, to each value of y in its range, there corresponds just one value of x; otherwise, the function is *multi-valued*. At times, it will be more convenient to label a given function of x as $f(x)$, to be read "the function of x" or "f of x." (Note carefully that this is not to be confused with "f times x.") If there are two functions, one may be labeled $f(x)$ and the other $g(x)$. Also, if $y = f(x) = x^2 - 5x + 4$, the statement "the value of the function is -2 when $x = 3$" can be replaced by "$f(3) = -2$."

Let $y = f(x)$. The set of values of the independent variable x is called the *domain* of the function while the range of the dependent variable is called the *range* of the function.

A variable w (dependent) is said to be a function of the (independent) variables $x, y, z, \ldots$ if, when a value of each of the variables $x, y, z, \ldots$ is known, there corresponds exactly one value of w. For example, the volume V of a rectangular parallelpiped of dimensions x, y, z is given by $V = xyz$. Here, V is a function of three independent variables.

Solved Problem 2-1. A piece of wire 30 in. long is bent to form a rectangle. If one of its dimensions is x in., express the area as a function of x.

Solution. Since the semiperimeter of the rectangle is $\frac{1}{2} \cdot 30 = 15$ in. and one dimension is x in., the other is $(15 - x)$ in. Thus, $A = x(15 - x)$.

Solved Problem 2-2. A right circular cylinder is said to be inscribed in a sphere if the circumferences of the bases of the cylinder are in the surface of the sphere. If the sphere has radius R, express the volume of the inscribed right circular cylinder as a function of the radius r of its base.

Solution. Let the altitude of the cylinder be denoted by $2h$. From Fig. 2-1, $h = \sqrt{R^2 - r^2}$ and the required volume is

$$V = \pi r^2 \cdot 2h = 2\pi r^2 \sqrt{R^2 - r^2}$$

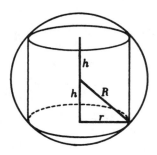

Figure 2-1

A function $y = f(x)$, by definition, yields a collection of number pairs $(x, f(x))$ or (x, y) in which x is any element in the domain of the function and $f(x)$ or y is the corresponding value of the function.

The rectangular Cartesian coordinate system in a plane is a device by which there is established a one-to-one correspondence between the points of the plane and ordered pairs of real numbers (a, b).

Consider two real number scales intersecting at right angles at O, the origin of each (see Fig. 2-2), and having the positive direction on the horizontal scale (now called the x *axis*) directed to the right and the positive direction on the vertical scale (now called the y *axis*) directed upward.

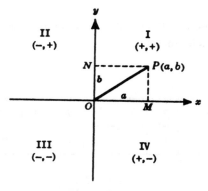

Figure 2-2

Let P be any point distinct from O in the plane of the two axes and join P to O by a straight line. Let the projection of OP on the x axis be $OM = a$ and the projection of OP on the y axis be $ON = b$. Then the pair of numbers (a, b) in that order are called the plane rectangular Cartesian coordinates (briefly, the rectangular coordinates) of P. In particular, the coordinates of O, the *origin* of the coordinate system, are $(0, 0)$.

The first coordinate, giving the directed distance of P from the y axis, is called the *abscissa* of P while the second coordinate, giving the directed distance of P from the x axis, is called the *ordinate* of P. Note carefully that the points $(3,4)$ and $(4,3)$ are distinct points.

The axes divide the plane into four sections, called *quadrants*. Fig. 2-2 shows the customary numbering of the quadrants and the respective signs of the coordinates of a point in each quadrant.

The graph of a function $f(x)$ consists of the totality of points (x, y) whose coordinates satisfy the relation $y = f(x)$. Any value of

x for which the corresponding value of a function $f(x)$ is zero is called a *zero* of the function. Such values of x are also called *roots* of the equation $f(x) = 0$. The real roots of an equation $f(x) = 0$ may be approximated by estimating from the graph of $f(x)$ the abscissas of its points of intersection with the x axis.

Solved Problem 2-3. Sketch the graph of the function $y = (x + 1)(x - 1)(x - 2)$. Refer to Table 2-2.

x	3	2	$\frac{3}{2}$	1	0	−1	−2
$y = f(x)$	8	0	$-\frac{5}{8}$	0	2	0	−12

Table 2-2

Solution. This is a *cubic* curve of the equation $y = (x + 1)(x - 1)(x - 2)$. It crosses the x axis where $x = -1$, 1, and 2. See Fig. 2-3.

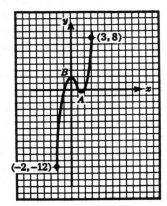

Figure 2-3

Solved Problem 2-4. Sketch the graph of the function $y = x^2 + 2x - 5$ and by means of it determine the real roots of $x^2 + 2x - 5 = 0$. Refer to Table 2-3.

Solution

x	2	1	0	−1	−2	−3	−4
$y = f(x)$	3	−2	−5	−6	−5	−2	3

Table 2-3

The parabola cuts the x axis at a point whose abscissa is between 1 and 2 (the value of the function changes sign) and at a point whose abscissa is between −3 and −4.

Reading from the graph in Fig. 2-4, the roots are $x = 1.5$ and $x = -3.5$, approximately.

Figure 2-4

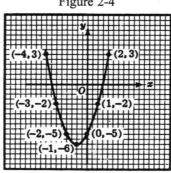

The Linear Equation

An equation is a statement that two expressions are equal. Examples of equations are (a) $2x - 6 = 4 - 3x$; (b) $y^2 + 3y = 4$; and (c) $2x + 3y = 4xy + 1$. An equation is linear in an unknown if the highest degree of that

unknown in the equation is one. An equation is quadratic in an unknown if the if the highest degree of that unknown is two.

Regarding the examples displayed above, the first is a *linear* equation in one unknown, the second is a *quadratic* in one unknown, and the third is linear in each of the two unknowns but is of degree two in the two unknowns. Any set of values of the unknowns for which the two members of an equation are equal is called a *solution* of the equation. Thus, $x = 2$ is a solution of (a) since $2(2) - 6 = 4 - 3(2)$; $y = 1$ and $y = -4$ are solutions of (b); and $x = 1, y = 1$ is a solution of (c).

A solution of an equation in one unknown is also called a *root* of the equation.

To solve a linear equation in one unknown, perform the same operations on both members of the equation in order to obtain the unknown alone in the left member.

Solved Problem 2-5. Solve $2x - 6 = 4 - 3x$.

Solution

Add 6: $\qquad$ $2x = 10 - 3x$

Add 3x: $\qquad$ $5x = 10$

Divide by 5: $\qquad$ $x = 2$

Check: $\qquad$ $2(2) - 6 = 4 - 3(2)$ or $-2 = -2$

Solved Problem 2-6. Solve $\dfrac{1}{3}x - \dfrac{1}{2} = \dfrac{3}{4}x + \dfrac{5}{6}$.

Solution

Multiply by LCD = 12: $\qquad$ $4x - 6 = 9x + 10$

Add $6-9x$:

$$-5x = 16$$

Divide by -5 :

$$x = -\frac{16}{5}$$

Check:

$$\frac{1}{3}\left(-\frac{16}{5}\right) - \frac{1}{2} = \frac{3}{4}\left(-\frac{16}{5}\right) + \frac{5}{6} \text{ or } -\frac{47}{30} = -\frac{47}{30}$$

An equation which contains fractions having the unknown in one or more denominators may sometimes reduce to a linear equation when cleared of fractions. When the resulting equation is solved, the solution *must* be checked since it may or may not be a root of the original equation.

Ratio and Proportion

The *ratio* of two quantities is their quotient. The ratio of 1 inch to 1 foot is 1/12 or 1:12, a pure number; the ratio of 30 miles to 45 minutes is 30/45 = 2/3 mile per minute. The expressed equality of two ratios, such as $\frac{a}{b} = \frac{c}{d}$ is called a *proportion*.

Variation

A variable y is said to vary *directly* as another variable x (or y is proportional to x) if y is equal to some constant c times x, that is, if $y = cx$.

A variable y is said to vary *inversely* as another variable x if y varies directly as the reciprocal of x, that is, if $y = c / x$.

A variable z is said to vary *jointly* as x and y if z varies directly as the product xy, that is, if $z = cxy$.

Simultaneous Linear Equations

Two Linear Equations in Two Unknowns

Let the system of equations be:

$$a_1x + b_1y + c_1 = 0$$
$$a_2x + b_2y + c_2 = 0$$

Each linear equation has an unlimited number of solutions (x, y) corresponding to the unlimited number of points on the locus (straight line) which it represents. Our problem is to find all solutions common to the two equations or the coordinates of all points common to the two lines. There are three cases:

(1) The system has one and only one solution, that is, the two lines have one and only one point in common. The equations are said to be *consistent* (have common solutions) and *independent*. See Fig. 2-5 indicating two distinct intersecting lines.

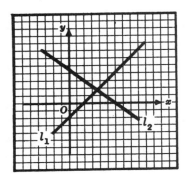

Figure 2-5

(2) The system has an unlimited number of solutions, that is, the two equations are equivalent or the two lines are coincident. The equations are said to be *consistent* and *dependent*. See Fig. 2-6 indicating that the two equations represent the same line.

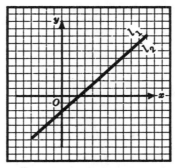

Figure 2-6

(3) The system has no solution, that is, the two lines are parallel and distinct. The equations are said to be *inconsistent*. See Fig. 2-7 indicating that the two equations result in two parallel lines.

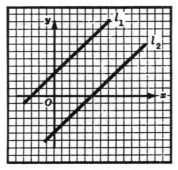

Figure 2-7

Solution of Simultaneous Linear Equations

Graphical Solution. We plot the graphs of the two equations on the same axes and scale off the coordinates of the point of intersection. The defect of this method is that, sometimes, only approximate solutions are obtained.

Solved Problem 2-7. Graphically solve the system

$$x + 2y = 5$$
$$3x - y = 1$$

Solution. The equations are graphed in Fig. 2-8.

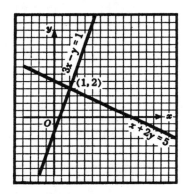

Figure 2-8

The solution of the system is $x = 1$, $y = 2$.

Algebraic Solution. A system of two consistent and independent equations in two unknowns may be solved algebraically by eliminating one of the unknowns.

Solved Problem 2-8. Algebraically solve the system

$$3x - 6y = 10 \qquad\qquad (2-1)$$
$$9x + 15y = -14 \qquad\qquad (2-2)$$

Solution

ELIMINATION BY SUBSTITUTION

Solve (2-1) for x:

$$x = \frac{10}{3} + 2y \qquad\qquad (2\text{-}3)$$

Substitute in (2-2):

$$9\left(\frac{10}{3} + 2y\right) + 15y = -14$$
$$30 + 18y + 15y = -14$$
$$33y = -44$$
$$y = -\frac{4}{3}$$

Substitute for y in (2-3):

$$x = \frac{10}{3} + 2\left(-\frac{4}{3}\right) = \frac{2}{3}$$

Check: Using (2-2),

$$9\left(\frac{2}{3}\right) + 15\left(-\frac{4}{3}\right) = -14$$

Solved Problem 2-9. Algebraically solve the system

$$2x - 3y = 10 \qquad\qquad (2\text{-}4)$$
$$3x - 4y = 8 \qquad\qquad (2\text{-}5)$$

Solution

ELIMINATION BY ADDITION

Multiply (2-4) by −3 and (2-5) by 2:

$$-6x + 9y = -30$$
$$\underline{6x - 8y = 16}$$
Add $\quad y = -14$

Substitute for x in (2-4):

$$2x + 42 = 10$$
$$x = -16$$

Check: Using (2-5),

$$3(-16) - 4(-14) = 8$$

Quadratic Functions and Equations

Quadratic Function

The graph of the quadratic equation, $y = ax^2 + bx + c, \ a \neq 0$, is a parabola. If $a > 0$, the parabola opens upward (Fig. 2-9); if $a < 0$, the parabola opens downward (Fig. 2-10). The lowest point of the parabola of Fig. 2-9 and the highest point of the parabola of Fig. 2-10 are called *vertices*. The abscissa of the vertex is given by $\qquad x = -\dfrac{b}{2a}$

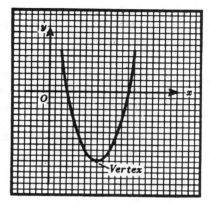

Figure 2-9

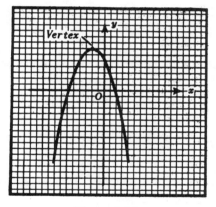

Figure 2-10

Quadratic Equation

A quadratic equation in one unknown x is of the form

$$ax^2 + bx + c = 0 \quad a \neq 0$$

Frequently, a quadratic equation may be solved by *factoring*.

Every quadratic equation can be solved by the following process, known as *completing the square*:

(a) Subtract the constant term c from both sides of the equation.
(b) Divide both sides of the equation by a, the coefficient of x^2.
(c) Add to each side the square of one-half the coefficient of the term in x.
(d) Set the square root of the left side (a perfect square) equal $\pm$ to the square root of the right side and solve for x.

Every quadratic equation can be solved by means of the quadratic formula:

$$x = \frac{-b \pm \sqrt{b^2 - 4ac}}{2a}$$

It should be noted that it is possible that the roots may be complex numbers.

Solved Problem 2-10. Solve $ax^2 + bx + c = 0$, $a \neq 0$ by completing the square.

Solution. Proceeding according to the four steps listed above:

(a) $\quad ax^2 + bx = -c$

(b) $\quad x^2 + \dfrac{b}{a}x = -\dfrac{c}{a}$

(c) $\quad x^2 + \dfrac{b}{a}x + \dfrac{b^2}{4a^2} = \dfrac{b^2}{4a^2} - \dfrac{c}{a} = \dfrac{b^2 - 4ac}{4a^2}$

(d) $\quad x + \dfrac{b}{2a} = \pm\sqrt{\dfrac{b^2 - 4ac}{4a^2}} = \pm\dfrac{\sqrt{b^2 - 4ac}}{2a}$

There, the solution to the quadratic equation is:

$$x = \frac{-b \pm \sqrt{b^2 - 4ac}}{2a}$$

Solved Problem 2-11. Solve the following equations using the quadratic formula: (a) $x^2 - 2x - 1 = 0$; (b) $3x^2 + 8x + 7 = 0$.

Solution

(a) $x = \dfrac{-(-2) \pm \sqrt{(-2)^2 - 4(1)(-1)}}{2 \cdot 1} = \dfrac{2 \pm 2\sqrt{2}}{2} = 1 \pm \sqrt{2}$

(b) $x = \dfrac{-(8) \pm \sqrt{8^2 - 4 \cdot 3 \cdot 7}}{2 \cdot 3} = \dfrac{-8 \pm \sqrt{-20}}{6} = \dfrac{-4 \pm i\sqrt{5}}{3}$

Discriminant of the Quadratic Equation

The discriminant of the quadratic equation is, by definition, the quantity $b^2 - 4ac$. When a, b, c are rational numbers, the roots of the equation are

real and unequal if and only if $b^2 - 4ac > 0$

real and equal if and only if $b^2 - 4ac = 0$

complex if and only if $b^2 - 4ac < 0$

Sum and Product of the Roots

If x_1 and x_2 are the roots of the quadratic equation, then $x_1 + x_2 = -\dfrac{b}{a}$ and $x_1 \cdot x_2 = \dfrac{c}{a}$.

A quadratic equation whose roots are x_1 and x_2 may be written in the form

$$x^2 - \left(x_1 + x_2\right)x + x_1 \cdot x_2 = 0$$

Simultaneous Equations Involving Quadratics

One Linear and One Quadratic Equation

Procedure: Solve the linear equation for one of the two unknowns (your choice) and substitute into the quadratic equation. Since this results in a quadratic equation of one unknown, the system can always be solved.

Solved Problem 2-12. Solve the system

$$4x^2 + 3y^2 = 16$$
$$5x + y = 7$$

Solution. Solve the linear equation for y:

$$y = 7 - 5x$$

Substitute in the quadratic equation:

$$4x^2 + 3(7-5x)^2 = 16$$
$$4x^2 + 3(49 - 70x + 25x^2) = 16$$
$$79x^2 - 210x + 131 = (x-1)(79x-131) = 0$$

and $x = 1, \dfrac{131}{79}$.

When $x = 1$, $y = 7 - 5x = 2$; when $x = \dfrac{131}{79}$, $y = -\dfrac{102}{79}$. The solu-

tions are $x = 1$, $y = 2$ and $x = \dfrac{131}{79}$, $y = -\dfrac{102}{79}$ (see Fig. 2-11).

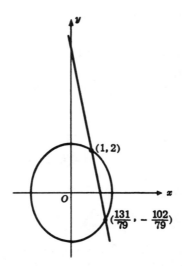

Figure 2-11

Two Quadratic Equations of the Form $ax^2 + by^2 = c$

Procedure: Eliminate one of the unknowns by the method of addition in the section on simultaneous linear equations.

Solved Problem 2-13. Solve the system

$$4x^2 + 9y^2 = 72 \tag{2-6}$$

$$3x^2 - 2y^2 = 19 \tag{2-7}$$

Solution

Multiply (2-6) by 2: $8x^2 + 18y^2 = 144$
Multiply (2-7) by 9: $27x^2 - 18y^2 = 171$
                Add: $35x^2 \qquad\quad = 315$

Then $x^2 = 9$ and $x = 3$.
When $x = 3$, (2-6) gives
$9y^2 = 72 - 4x^2 = 72 - 36 = 36,\ y^2 = 4,\ \text{and } y = \pm 2$.

When $x = -3$, (2-6) gives
$9y^2 = 72 - 4x^2 = 72 - 36 = 36,\ y^2 = 4,\ \text{and } y = \pm 2$.

The four solutions $x = 3,\ y = 2;\ x = 3,\ y = -2;\ x = -3,\ y = 2;\ x = -3,\ y = -2$ may also be written as $x = \pm 3,\ y = \pm 2;\ x = \pm 3,\ y = \mp 2$.

The ellipse and the hyperbola intersect in the points (3,2), (3,−2), (−3,2), (−3,−2). See Fig. 2-12.

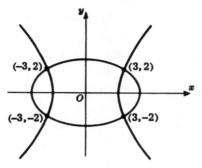

Figure 2-12

Two Quadratic Equations, One Homogeneous

An expression, as $2x^2 - 3xy + y^2$, whose terms are all of the same degree in the variables, is called *homogeneous*. A homogeneous expression equated to zero is called a *homogeneous equation*. A homogeneous quadratic equation in two unknowns can always be solved for one of the unknowns in terms of the other.

Solved Problem 2-14. Solve the system

$$x^2 - 3xy + 2y^2 = 0 \tag{2-8}$$

$$2x^2 + 3xy - y^2 = 13 \tag{2-9}$$

Solution. Solve (2-8) for x in terms of y:

$$(x - y)(x - 2y) = 0 \text{ and } x = y, \ x = 2y$$

Solve the systems:

$2x^2 + 3xy - y^2 = 13$ $2x^2 + 3xy - y^2 = 13$

$x = y$ $x = 2y$

$2y^2 + 3y^2 - y^2 = 4y^2 = 13$ $8y^2 + 6y^2 - y^2 = 13y^2 = 13$

$y^2 = \dfrac{13}{4}, \; y = \pm\dfrac{\sqrt{13}}{2}$ $y^2 = 1, \; y = \pm 1$

Then $x = y = \pm\dfrac{\sqrt{13}}{2}$ Then $x = 2y = \pm 2$

The solutions are $x = \dfrac{\sqrt{13}}{2}, \; y = \dfrac{\sqrt{13}}{2}$; $x = -\dfrac{\sqrt{13}}{2}, \; y = -\dfrac{\sqrt{13}}{2}; x = 2,$

$y = 1; x = -2, y = -1$ or $x = \pm\dfrac{\sqrt{13}}{2}, \; y = \pm\dfrac{\sqrt{13}}{2}; x = \pm 2, \; y = \pm 1$.

Two Quadratic Equations of the Form
$ax^2 + bxy + cy^2 = d$

Procedure: Combine the two given equations to obtain a homogeneous equation. Solve the system consisting of this
homogeneous equation and either of the given equations.

Two Quadratic Equations,
Each Symmetrical in x and y

Procedure: Substitute $x = u + v$ and $y = u - v$ and then eliminate
v^2 from the resulting equations.

Polynomial Equations, Rational Roots

A *polynomial equation* (or rational integral equation) is obtained when any polynomial in one variable is set equal to zero. A polynomial equation is said to be in *standard form* when written as

$$a_o x^n + a_1 x^{n-1} + a_2 x^{n-2} + \cdots + a_{n-2} x^2 + a_{n-1} x + a_n = 0$$

where the terms are arranged in descending powers of x. A zero has been inserted as coefficient of each missing term. The coefficients have no common factor except ± 1, and $a_0 \neq 0$.

A number r is called a *root* of $f(x) = 0$ if and only if $f(r) = 0$. It follows that the abscissas of the points of intersection of the graph $y = f(x)$ and the x axis are roots of $f(x) = 0$.

The Fundamental Theorem of Algebra

Every polynomial equation $f(x) = 0$ has at least one root, real or complex.

A polynomial equation of degree n has exactly n roots. These n roots may not all be distinct. If r is one of the roots and occurs just once, it is called a *simple root*; if r occurs exactly $m > 1$ times among the roots, it is called a *root of multiplicity m* or an *m–fold root*. If $m = 2$, r is called a *double root*; if $m = 3$, a *triple root*; and so on.

Solved Problem 2-15. Find all of the roots of $(x+1)(x-2)^3 (x+4)^2 = 0$.

Solution. The roots are -1, 2, 2, 2, -4, -4; thus, -1 is a simple root, 2 is a root of multiplicity three or a triple root, and -4 is a root of multiplicity two or a double root.

Complex Roots

If the polynomial equation $f(x) = 0$ has real coefficients and if the complex $a + bi$ is a root of $f(x) = 0$, then the *complex conjugate* $a - bi$ is also a root.

Irrational Roots

Given the polynomial equation $f(x) = 0$, if the irrational number $a + \sqrt{b}$, where a and b are rational, is a root of $f(x) = 0$, then the conjugate irrational $a - \sqrt{b}$ is also a root.

Limits to the Real Roots

A real number L is called an *upper limit* of the real roots of $f(x) = 0$ if no (real) root is greater than L; a real number l is called a *lower limit* if no (real) root is smaller than l.

If $L > 0$ and if, when $f(x)$ is divided by $x - L$ by synthetic division, every number in the third line is nonnegative, then L is an upper limit of the real roots of $f(x) = 0$.

If $l < 0$ and if, when $f(x)$ is divided by $x - l$ by synthetic division, the numbers in the third line alternate in sign, then l is a lower limit of the real roots of $f(x) = 0$.

Rational Roots

A polynomial equation has 0 as a root if and only if the constant term of the equation is zero.

If a rational fraction p/q, expressed in lowest terms, is a root of the polynomial equation in which $a_n \neq 0$, then p is a divisor of the constant term a_n and q is a divisor of the leading coefficient of a_0.

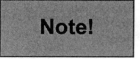

If p, an integer, is a root of the polynomial equation, then p is a divisor of its constant term.

The principal problem of this section is to find the rational roots of a given polynomial equation. The general procedure is: Test the possible rational roots by synthetic division, accepting as roots all those for which the last number in the third line *is* zero and rejecting all those for which it is not.

Solved Problem 2-16. Find the rational roots of

$x^5 + 2x^4 - 18x^3 - 8x^2 + 41x + 30 = 0$.

Solution. Since the leading coefficient is 1, all rational roots p/q are integers. The possible integral roots, the divisors (both positive and negative) of the constant term 30, are:

$\pm 1, \pm 2, \pm 3, \pm 5, \pm 6, \pm 10, \pm 15, \pm 30$

$$
\begin{array}{r}
1 + 2 - 18 - \ 8 + 41 + 30 \quad \underline{|1} \\
\text{Try 1:} \quad \underline{1 + \ 3 - 15 - 23 + 18} \\
1 + 3 - 15 - 23 + 18 + 48
\end{array}
$$

Then 1 is not a root. This number (+1) should be removed from the list of possible roots lest we forget and try it again later on.

$$
\begin{array}{r}
1 + 2 - 18 - \ 8 + 41 + 30 \quad \underline{|2} \\
\text{Try 2:} \quad \underline{2 + 8 - 20 - 56 - 30} \\
1 + 4 - 10 - 28 - 15 + \ 0
\end{array}
$$

Then 2 is a root and the remaining rational roots of the given equation are the rational roots of the *depressed equation*:

$x^4 + 4x^3 - 10x^2 - 28x - 15 = 0$

Now ±2, ±6, ±10, and ±30 cannot be roots of this equation (they are not divisors of 15) and should be removed from the list of possibilities. We return to the depressed equation:

$$
\begin{array}{r}
1+4-10-28-15 \quad \underline{|3} \\
\text{Try 3:} \qquad \underline{3+21+33+15} \\
1+7+11+\ 5+\ 0
\end{array}
$$

Then 3 is a root and the new depressed equation is:

$$x^3 + 7x^2 + 11x + 5 = 0$$

Since the coefficients of this equation are nonnegative, it has no positive roots. We now remove ±3, ±5, ±15 from the original list of possible roots and return to the new depressed equation:

$$
\begin{array}{r}
1+7+11+5 \quad \underline{|-1} \\
\text{Try } -1: \qquad \underline{-1-\ 6-5} \\
1+6+\ 5+0
\end{array}
$$

Then -1 is a root and the depressed equation

$$x^2 + 6x + 5 = (x+1)(x+5) = 0$$

has -1 and -5 as roots.

The necessary computations may be neatly displayed as follows:

$$
\begin{array}{ll}
1+2-18-\ 8+41+30 & \underline{|2} \\
\quad 2+8-20-56-30 & \\
1+4-10-28-15 & \underline{|3} \\
\quad 3+21+33+15 & \\
1+7+11+\ 5 & \underline{|-1} \\
\quad -1-6-\ 5 & \\
1+6+\ 5 & \\
x^2+6x+5=(x+1)(x+5)=0 & \quad x=-1,-5
\end{array}
$$

The roots are 2, 3, -1, -1, -5.

Note that the roots here are numerically small numbers, that is, 3 is a root but 30 is not, -1 is a root but -15 is not.

Graphs of Polynomials

The general polynomial (or rational integral function) of the nth degree in x has the form

$$f(x)=a_ox^n+a_1x^{n-1}+a_2x^{n-2}+\cdots+a_{n-2}x^2+a_{n-1}x+a_n$$

in which n is a positive integer and the a's are constants, real or complex, with $a_o \neq 0$. The term a_ox_n is called the *leading term*, a_n the *constant term*, and a_o the *leading coefficient*.

Remainder Theorem

If a polynomial $f(x)$ is divided by $x-h$ until a remainder free of x is obtained, this remainder is $f(h)$.

Example 2-1. Let $f(x) = x^3 + 2x^2 - 3x - 4$ and $x - h = x - 2$; then $h = 2$. By actual division:

$$\frac{x^3 + 2x^2 - 3x - 4}{x - 2} = x^2 + 4x + 5 + \frac{6}{x - 2}$$

or $x^3 + 2x^2 - 3x - 4 = (x^2 + 4x + 5)(x - 2) + 6$, and the remainder is 6.

By the remainder theorem, the remainder is

$$f(2) = 2^3 + 2 \cdot 2^2 - 3 \cdot 2 - 4 = 6$$

Factor Theorem

If $x - h$ is a factor of $f(x)$, then $f(h) = 0$, and conversely.

Synthetic Division

By a process known as *synthetic division*, the necessary work in dividing a polynomial $f(x)$ by $x - h$ may be displayed in three lines, as follows:

(1) Arrange the dividend $f(x)$ in descending powers of x (as usual in division) and write down in the first line the coefficients, supplying zero as coefficient whenever a term is missing.

(2) Place h, the synthetic divisor, in the first line to the right of the coefficients.

(3) Recopy the leading coefficient a_{ns} directly below it in the third line.

(4) Multiply a_o by h; place the product $a_o h$ in the second line under a_1 (in the first line), add to a_1, and place the sum $a_o h + a_1$ in the third line under a_1.

(5) Multiply the sum in Step 4 by h; place the product in the second line under a_2, add to a_2, and place the sum in the third line under a_2.

(6) Repeat the process of Step 5 until a product has been added to the constant term a_n.

The first n numbers in the third line are the coefficients of the quotient, a polynomial of degree $n - 1$, and the last number of the third line is the remainder $f(h)$.

Solved Problem 2-17. Divide $5x^4 - 8x^2 - 15x - 6$ by $x - 2$, using synthetic division.

Solution. Following the procedure outlined above, we have

$$
\begin{array}{l}
5 \;+0-\; 8 \;-15-\; 6 \quad \underline{|2} \\
 10+\; 20 + 24 + 18 \\
\hline
5 + 10 + 12 +\; 9 +\; 12
\end{array}
$$

The quotient is

$$Q(x) = 5x^3 + 10x^2 + 12x + 9$$

and the remainder is $f(2) = 12$.

Descartes' Rule of Signs

The number of positive roots of a polynomial equation $f(x) = 0$, with real coefficients, is equal either to the number of variations of sign in $f(x)$ or to that number diminished by an even number.

The number of negative roots of $f(x) = 0$ is equal to the number of positive roots of $f(-x) = 0$.

Example 2-2. Consider the polynomial $P(-x) = 2x^2 + x - 4$. $P(x)$ has *one* variation of sign (from + to −), $P(-x) = 2x^2 - x - 4$ has *one* variation of sign (from + to −). Descartes' rule of signs tells us that there will be one positive zero [one variation in sign for $P(x)$] and one negative zero [one variation in sign for $P(-x)$].

Graph of a Polynomial

The graph of a polynomial $y = f(x)$ may be obtained by computing a table of values, locating the several points (x, y), and joining them by a smooth curve. In order to avoid unnecessary labor in constructing the table, the following systematic procedure is suggested:

(1) When $x = 0$, $y = f(0)$ is the constant term of the polynomial.

(2) Use synthetic division to find $f(1), f(2), f(3), \ldots$ stopping as soon as the numbers in the third line of the synthetic division have the same sign.

(3) Use synthetic division to find $f(-1), f(-2), f(-3), \ldots$ stopping as soon as the numbers in the third line of the synthetic division have alternating signs.

In advanced mathematics, it is proved:

(a) The graph of a polynomial in x with integral coefficients is always a smooth curve without breaks or sharp corners.

(b) The number of *real* intersections of the graph of a polynomial of degree n with the x axis is *never* greater than n.

(c) If a and b are real numbers such that $f(a)$ and $f(b)$ have opposite signs, the graph has an *odd* number of real intersections with the x axis between $x = a$ and $x = b$.

(d) If a and b are real numbers such that $f(a)$ and $f(b)$ have the same signs, the graph either does not intersect the x axis or intersects it an *even* number of times between $x = a$ and $x = b$. See

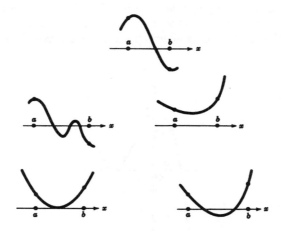

Fig. 2-13.

Figure 2-13

Solved Problem 2-18. Sketch the graph of $y = f(x) = x^3 - 4x^2 - 3x + 18$.

x	−3	−2	−1	0	1	2	3	4	5
y	−36	0	16	18	12	4	0	6	28

Table 2-4

Solution. From Table 2-4, it is evident that the graph crosses the x axis at $x = -2$ and meets it again at $x = 3$. If there is a third distinct intersection, it must be between $x = 2$ and $x = 3$ or between $x = 3$ and $x = 4$. By computing additional points for x on these intervals, we are led to suspect that no such third intersection exists.

This function has been selected so that the question of intersections can be definitely settled. When $f(x)$ is divided by $x + 2$, the quotient is $x^2 - 6x + 9 = (x - 3)^2$ and the remainder is zero. Thus, $f(x) = (x + 2)(x - 3)^2$ in factored form. It is now clear that the function is positive for $x > -2$, that is, the graph never falls below the x axis on this interval. Thus, the graph is tangent to the x axis at $x = 3$, the point of tangency accounting for two of its intersections with the x axis. (See Fig. 2-14).

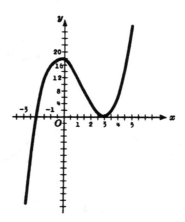

Figure 2-14

Partial Fractions

A rational algebraic fraction is the quotient of two polynomials. A rational fraction is called *proper* if the degree of the numerator is less than that of the denominator; otherwise, the fraction is called *improper*.

An improper fraction may be written as the sum of a polynomial and a proper fraction. For example, $\dfrac{3}{x+2}$ is a proper fraction while $\dfrac{x^2+3x+5}{x+2}$ is an improper fraction. Note that $\dfrac{x^2+3x+5}{x+2}=x+1+\dfrac{3}{x+2}$ is the sum of a polynomial and a proper fraction.

Two or more proper fractions may be summed to yield a single fraction whose denominator is the lowest common denominator of the several fractions. For example,

(a) $\dfrac{1}{x+2}+\dfrac{2}{3x-2}=\dfrac{5x+2}{(x+2)(3x-2)}$

(b) $\dfrac{3}{x+2}+\dfrac{4}{(x+2)^2}-\dfrac{1}{x}=\dfrac{2x^2+6x-4}{x(x+2)^2}$

(c) $\dfrac{1}{x-1}+\dfrac{3x-1}{x^2+2}-\dfrac{1}{(x^2+2)^2}=\dfrac{4x^4-4x^3+11x^2-9x+7}{(x-1)(x^2+2)^2}$

The problem of this section is to reverse the above process, that is, to resolve a given rational fraction into a sum of simpler proper fractions, called *partial fractions*.

CASE 1. Factors of the Denominator Linear, None Repeated

Corresponding to each factor of the denominator form a partial fraction having an unknown constant as numerator and the factor as denominator.

Solved Problem 2-19. Resolve $\dfrac{5x+2}{(x+2)(3x-2)}$ into partial fractions.

Solution. Set

$$\frac{5x+2}{(x+2)(3x-2)} = \frac{A}{x+2} + \frac{B}{3x-2} \text{ . Then,}$$

(a) $\dfrac{5x+2}{(x+2)(3x-2)} = \dfrac{A(3x-2)+B(x+2)}{(x+2)(3x-2)}$ and

(b) $\dfrac{5x+2}{(x+2)(3x-2)} = \dfrac{(3A+B)x-2A+2B}{(x+2)(3x-2)}$

are identities which hold for all values of x except possibly for $x = -2$ and $x = \dfrac{2}{3}$.

Solution 1. Equating coefficients of like term in the two members of the identity $5x + 2 = (3A + B)x - 2A + 2B$, we have $3A + B = 5$, $-2A + 2B = 2$. Then $A = I$, $B = 2$, and

$$\frac{5x+2}{(x+2)(3x-2)} = \frac{1}{x+2} + \frac{2}{3x-2}$$

Solution 2. Consider the identity $5x + 2 = A(3x - 2) + B(x + 2)$. Since it is an identity between polynomials, it holds for *all* values of x. Now when

$x = -2$, coefficient of B is 0, $5(-2) + 2 = A[3(-2) - 2]$, and $A = 1$;

$x = \dfrac{2}{3}$, coefficient of A is 0, $5\left(\dfrac{2}{3}\right) + 2 = B\left(\dfrac{2}{3} + 2\right)$, and $B = 2$.

Thus, as before,

$$\frac{5x+2}{(x+2)(3x-2)} = \frac{1}{x+2} + \frac{2}{3x-2}$$

CASE 2. Factors of the Denominator Linear, Some Repeated

Here (B) suggests that for each repeated factor $(ax + b)^k$, we set up a series of partial fractions

$$\frac{A}{(ax+b)^k} + \frac{B}{(ax+b)^{k-1}} + \cdots + \frac{K}{ax+b}$$

Solved Problem 2-20. Resolve into partial fractions:

$$\frac{2x^2 + 6x - 4}{x(x+2)^2}$$

Solution. Set

$$\frac{2x^2 + 6x - 4}{x(x+2)^2} = \frac{A}{(x+2)^2} + \frac{B}{x+2} + \frac{C}{x}$$

$$= \frac{Ax + Bx(x+2) + C(x+2)^2}{x(x+2)^2}$$

$$= \frac{(B+C)x^2 + (A+2B+4C)x + 4C}{x(x+2)^2}$$

Solution 1. From the identity $2x^2 + 6x - 4 = (B + C)x^2 + (A + 2B + 4C)x + 4C$; by equating coefficients of like powers of x, we obtain

$B + C = 2$, $A + 2B + 4C = 6$, $4C = -4$. Then $C = -1$, $B = 2 - C = 3$, $A = 6 - 2B - 4C = 4$, and

$$\frac{2x^2 + 6x - 4}{x(x+2)^2} = \frac{4}{(x+2)^2} + \frac{3}{x+2} - \frac{1}{x}$$

Solution 2. Consider the identity $2x^2 + 6x - 4 = Ax + Bx(x + 2) + C(x + 2)^2$. Using $x = -2$, we find $A = 4$; using $x = 0$, we find $C = -1$; using $x = 1$, we have $2 + 6 - 4 = A + 3B + 9C = 4 + 3B - 9$ and $B = 3$. These are the values of A, B, C found above.

Note that only two values $x = -2$ and $x = 0$ are suggested by the identity. Since three constants are to be determined, one additional value of x is needed. It may be taken at random.

CASE 3. Denominator Contains Irreducible Quadratic Factors, None Repeated

For each irreducible quadratic factor $ax^2 + bx + c$ of the denominator set up a partial fraction of the form $\dfrac{Ax + B}{ax^2 + bx + c}$.

Solved Problem 2-21. Resolve $\dfrac{2}{(x-1)(x^2 + x - 4)}$ into partial fractions.

Solution. Set

$$\frac{2}{(x-1)(x^2+x-4)} = \frac{Ax+B}{x^2+x-4} + \frac{C}{x-1}$$

$$= \frac{(Ax+B)(x-1)+C(x^2+x-4)}{(x-1)(x^2+x-4)}$$

$$= \frac{(A+C)x^2+(B-A+C)x-B-4C}{(x-1)(x^2+x-4)}$$

From the identity $2 = (A + C)x^2 + (B - A + C)x - B - 4C$, we have

$$A + C = 0 \qquad\qquad B - A + C = 0 \qquad\qquad -B - 4C = 2$$

with solution $A = 1$, $B = 2$, $C = -1$. Then

$$\frac{2}{(x-1)(x^2+x-4)} = \frac{x+2}{x^2+x-4} - \frac{1}{x-1}$$

CASE 4. Denominator Contains Irreducible Quadratic Factors, Some Repeated

Here (C) suggests that for each repeated irreducible factor $(ax^2+bx+c)^k$ of the denominator, we set up a series of partial fractions

$$\frac{Ax+B}{(ax^2+bx+c)^k} + \frac{Cx+D}{(ax^2+bx+c)^{k-1}} + \cdots + \frac{Hx+K}{ax^2+bx+c}$$

Solved Problem 2-22. Resolve $\dfrac{x^4-x^3+8x^2-6x+7}{(x-1)(x^2+2)^2}$ into partial fractions.

Solution. Set

$$\frac{x^4 - x^3 + 8x^2 - 6x + 7}{(x-1)(x^2+2)^2} = \frac{A}{x-1} + \frac{Bx+C}{(x^2+2)^2} + \frac{Dx+E}{x^2+2}$$

$$= \frac{A(x^2+2)^2 + (Bx+C)(x-1) + (Dx+E)(x^2+2)(x-1)}{(x-1)(x^2+2)^2}$$

$$= \frac{(A+D)x^4 + (-D+E)x^3 + (4A+B+2D-E)x^2}{(x-1)(x^2+2)^2} +$$

$$\frac{(-B+C-2D+2E)x + 4A-C-2E}{(x-1)(x^2+2)^2}$$

Equating coefficients of like terms in the identity

$$x^4 - x^3 + 8x^2 - 6x + 7 = (A+D)x^4 + (-D+E)x^3 +$$
$$(4A+B+2D-E)x^2 + (-B+C-2D+2E)x + 4A-C-2E,$$

we have

$$A + D = 1$$
$$-D + E = -1$$
$$4A + B + 2D - E = 8$$
$$-B + C - 2D + 2E = -6$$
$$4A - C - 2E = 7$$

From the identity,

$$x^4 - x^3 + 8x^2 - 6x + 7 = A\left(x^2 + 2\right)^2 + (Bx + C)(x - 1) +$$
$$(Dx + E)\left(x^2 + 2\right)(x - 1)$$

Using $x = 1$, we find $A = 1$. Then

$D = 1 - A = 0$
$E = -1 + D = -1$
$B = 3$
$C = -1$

and

$$\frac{x^4 - x^3 + 8x^2 - 6x + 7}{(x - 1)\left(x^2 + 2\right)^2} = \frac{1}{x - 1} + \frac{3x - 1}{\left(x^2 + 2\right)^2} - \frac{1}{x^2 + 2}$$

Chapter 3
PROGRESSIONS, SEQUENCES, AND

IN THIS CHAPTER:

✔ *Arithmetic and Geometric Progressions*
✔ *Infinite Geometric Series*
✔ *Mathematical Induction*
✔ *The Binomial Theorem*
✔ *Infinite Sequences*
✔ *Infinite Series*
✔ *Power Series*

Arithmetic and Geometric Progressions

Sequences

A *sequence* is a set of numbers, called *terms*, arranged in a definite order; that is, there is a rule by which the terms after the first may be formed. Sequences may be finite or infinite.

Example 3-1.

(a) Sequence: 3, 7, 11, 15, 19, 23, 27
     Type: Finite of 7 terms.
     Rule: Add 4 to a given term to produce the next.

(b) Sequence: 3, 6, 12, 24, 48, 96
     Type: Finite of 6 terms.
     Rule: Multiply a given term by 2 to produce the next.

Arithmetic Progressions

An *arithmetic progression* is a sequence in which each term after the first is formed by adding a fixed amount, called the *common difference*, to the preceding term. The sequence of Example 3-1(a) is an arithmetic progression whose common difference is 4.

If a is the first term, d is the common difference, and n is the number of terms of an arithmetic progression, the successive terms are:

$$a, a + d, a + 2d, a + 3d, ..., a + (n - 1)d$$

Thus, the *last* term (or nth term) l is given by

$$l = a + (n - 1)d$$

The *sum S* of the n terms of this progression is given by

$$S = \frac{n}{2}(a + l) \quad \text{or} \quad S = \frac{n}{2}\left[2a + (n - 1)d\right]$$

Solved Problem 3-1. Find the twentieth term and the sum of the first 20 terms of the arithmetic progression 4, 9, 14, 19, ...

Solution. For this progression, $a = 4$, $d = 5$, and $n = 20$; then the twentieth term is

$$l = a + (n-1)d = 4 + 19 \cdot 5 = 99$$

and the sum of the first 20 terms is

$$S = \frac{n}{2}(a+l) = \frac{20}{2}(4+99) = 1030$$

Solved Problem 3-2. Find the value of k such that each of the following sequences is an arithmetic progression: (a) $k - 1$, $k + 3$, $3k - 1$; (b) $3k^2 + k + 1$, $2k^2 + k$, $4k^2 - 6k + 1$.

Solution.

(a) If the sequence is to form an arithmetic progression,

$$(k+3) - (k-1) = (3k-1) - (k+3)$$

Then, $k = 4$ and the arithmetic progression is 3, 7, 11.

(b) Setting

$$(2k^2 + k) - (3k^2 + k + 1) = (4k^2 - 6k + 1) - (2k^2 + k)$$

we have $\qquad 3k^2 - 7k + 2 = 0$

and $\qquad k = 2, \frac{1}{3}.$

The progressions are 15, 10, 5 when $k = 2$ and $\frac{5}{3}, \frac{5}{9}, -\frac{5}{9}$ when $k = \frac{1}{3}$.

The terms between the first and last terms of an arithmetic progression are called *arithmetic means* between these two terms. Thus, to insert k arithmetic means between two numbers is to form an arithmetic progression of $(k + 2)$ terms having the two given numbers as first and last terms.

Solved Problem 3-3. Insert five arithmetic means between 4 and 22.

Solution. We have $a = 4$, $l = 22$, and $n = 5 + 2 = 7$. Then, $22 = 4 + 6d$ and $d = 3$. The first mean is $4 + 3 = 7$, the second is $7 + 3 = 10$, and so on. The required means are 7, 10, 13, 16, 19, and the resulting progression is 4, 7, 10, 13, 16, 19, 22.

When just one mean is to be inserted between two numbers, it is called the *arithmetic mean* (also, the average) of the two numbers.

Solved Problem 3-4. Find the arithmetic mean of the two numbers a and l.

Solution. We seek the middle term of an arithmetic progression of three terms having a and l as first and third terms, respectively. If d is the common difference, then $a + d = l - d$ and $d = \frac{1}{2}(l - a)$. The arithmetic mean is

$$a + d = a + \frac{1}{2}(l - a) = \frac{1}{2}(a + l)$$

Geometric Progressions

A *geometric progression* is a sequence in which each term after the first is formed by multiplying the preceding term by a fixed number, called the *common ratio*. The sequence 3, 6, 12, 24, 48, 96 of Example 3-1(b) is a geometric progression whose common ratio is 2.

If a is the first term, r is the common ratio, and n is the number of terms, the geometric progression is

$$a, ar, ar^2, ..., ar^{n-1}$$

Thus, the last (or nth) term l is given by

$$l = ar^{n-1}$$

The *sum* S of the first n terms of the geometric progression $a, ar, ar^2, ..., ar^{n-1}$ is given by

$$S = \frac{a - rl}{1 - r} \quad \text{or} \quad S = \frac{a(1 - r^n)}{1 - r}$$

Solved Problem 3-5. Find the ninth term and the sum of the first nine terms of the geometric progression 8, 4, 2, 1....

Solution. Here $a = 8$, $r = \dfrac{1}{2}$, and $n = 9$; the ninth term is

$$l = ar^{n-1} = 8\left(\frac{1}{2}\right)^8 = \left(\frac{1}{2}\right)^5 = \frac{1}{32}$$

and the sum of the first nine terms is

$$S = \frac{a - rl}{1 - r} = \frac{8 - \dfrac{1}{2}\left(\dfrac{1}{32}\right)}{1 - \dfrac{1}{2}} = 16 - \frac{1}{32} = \frac{511}{32}$$

The terms between the first and last terms of a geometric progression are called *geometric means* between the two terms. Thus, to insert k geometric means between two numbers is to form a geometric progression of $(k + 2)$ terms having the two given numbers as first and last terms.

Solved Problem 3-6. Insert four geometric means between 25 and $\dfrac{1}{125}$.

Solution. We have $a = 25$, $l = \dfrac{1}{125}$, and $n = 4 + 2 = 6$. Using $l = ar^{n-1}$,

$\dfrac{1}{125} = 25r^5$; then $r^5 = \left(\dfrac{1}{5}\right)^5$ and $r = \dfrac{1}{5}$. The first mean is

$25\left(\dfrac{1}{5}\right) = 5$, the second is $5\left(\dfrac{1}{5}\right) = 1$, and so on. The required means

are $5, 1, \dfrac{1}{5}, \dfrac{1}{25}$ and the geometric progression is $25, 5, 1, \dfrac{1}{5}, \dfrac{1}{25}, \dfrac{1}{125}$.

When one mean is to be inserted between two numbers, it is called the *geometric mean* of the two numbers. The geometric mean of two numbers a and l, having like signs, is $(\pm)\sqrt{a \cdot l}$. The sign to be used is the common sign of a and l.

Infinite Geometric Series

The indicated sum of the terms of a finite or infinite sequence is called a finite or infinite *series*. The sums of arithmetic and geometric progressions in the preceding section are examples of finite series.

Of course, it is impossible to add up all the terms of an infinite series; that is, in the usual meaning of the word *sum*, there is no such thing as the sum of such a series. However, it is possible to associate with certain infinite series, a well-defined number which, for convenience, will be called the sum of the series.

Example 3-2. From the table of values of $\left(\dfrac{1}{2}\right)^n$ in Table 3-1, it appears

that, as n increases indefinitely, $\left(\dfrac{1}{2}\right)^n$ decreases indefinitely while

remaining positive. Moreover, it can be made to have a value as near 0 as we please by choosing n sufficiently large. We describe this state of

affairs by saying: The limit of $\left(\dfrac{1}{2}\right)^n$, as n increases indefinitely, is 0.

n	1	3	5	10
$(\frac{1}{2})^n$	0.5	0.125	0.03125	0.0009765625

Table 3-1

By examining the behavior of r^n for other values of r, it becomes clear that:

The limit of r^n, as n increases indefinitely, is 0 when $|r| < 1$.

The sum, S, of the infinite geometric series

$$a + ar + ar^2 + \cdots + ar^{n-1} + \cdots, \quad |r| < 1, \quad \text{is } S = \frac{a}{1-r} \quad .$$

Example 3-3. For the infinite geometric series

$$12 + 4 + \frac{4}{3} + \frac{4}{9} + \cdots, \quad a = 12 \text{ and } r = \frac{1}{3} \quad . \text{ The sum of the series is}$$

$$S = \frac{a}{1-r} = \frac{12}{1-\frac{1}{3}} = 18 \quad .$$

Every repeating decimal approximates a rational number. This rational number is also called the *limiting value of the decimal*.

Example 3-4. Find the limiting value of the repeating decimal 0.727272…. We write 0.727272… = 0.72 + 0.0072 + 0.000072 + … and note that for this infinite geometric series $a = 0.72$ and $r = 0.01$. Then

$$S = \frac{a}{1-r} = \frac{0.72}{1-0.01} = \frac{0.72}{0.99} = \frac{72}{99} = \frac{8}{11}$$

Sigma Notation

Sigma notation is a convenient notation for expressing sums. $\sum_{i=a_1}^{a_m} x_i$ simply means: add $x_{a_1} + x_{a_2} + \cdots + x_{a_{m-1}} + x_m$. For example,

$$\sum_{i=1}^{4} 2i = 2(1) + 2(2) + 2(3) + 2(4) = 2 + 4 + 6 + 8 = 20$$

$$\sum_{j=2}^{6} j^2 = 2^2 + 3^2 + 4^2 + 5^2 + 6^2 = 4 + 9 + 16 + 25 + 36 = 90$$

Thus,

$$a + ar + ar^2 + \cdots + ar^{n-1} + \cdots = \frac{a}{1-r} = \sum_{i=0}^{\infty} ar^i, \quad |r| < 1$$

Solved Problem 3-7. A rubber ball is dropped from a height of 81 m. Each time it strikes the ground, it rebounds two-thirds of the distance through which it last fell. Find the total distance it travels in coming to rest.

Solution. For the falls: $a = 81$ and $r = \dfrac{2}{3}$; $S = \dfrac{a}{1-r} = \dfrac{81}{1 - \dfrac{2}{3}} = 243$ m.

For the rebounds: $a = 54$ and $r = \dfrac{2}{3}$; $S = \dfrac{54}{1 - \dfrac{2}{3}} = 162$ m.

Thus, the total distance traveled is 243 m + 162 m = 405 m.

Mathematical Induction

Everyone is familiar with the process of reasoning, called *ordinary* or *incomplete induction*, in which a generalization is made on the basis of a number of simple observations.

Example 3-5. We observe that $1 = 1^2$, $1 + 3 = 4 = 2^2$, $1 + 3 + 5 = 9 = 3^2$, $1 + 3 + 5 + 7 = 16 = 4^2$, and conclude that

$$1 + 3 + 5 + \cdots + (2n - 1) = n^2$$

or, in words, the sum of the first n odd integers is n^2.

Example 3-6. We observe that 2 points determine $1 = \dfrac{1}{2} \cdot 2(2 - 1)$ line; that 3 points, not on a line, determine $3 = \dfrac{1}{2} \cdot 3(3 - 1)$ lines; that 4 points, no 3 on a line, determine $6 = \dfrac{1}{2} \cdot 4(4 - 1)$ lines; that 5 points, no 3 on a line, determine $10 = \dfrac{1}{2} \cdot 5(5 - 1)$ lines; and conclude that n points, no 3 on a line, determine $\dfrac{1}{2} n(n - 1)$ lines.

Example 3-7. We observe that, for $n = 1, 2, 3, 4, 5$, the values of

$$f(n) = \frac{n^4}{8} - \frac{17n^3}{12} + \frac{47n^2}{8} - \frac{103n}{12} + 6$$

are 2, 3, 5, 7, 11, respectively, and conclude that $f(n)$ is a prime number for every positive integral value of n.

The conclusions in Examples 3-5 and 3-6 are valid. The conclusions in Example 3-7 is false since $f(6) = 22$ is not a prime number.

Complete Induction

Mathematical induction or complete induction is a type of reasoning by which such conclusions as were drawn in the above examples may be proved or disproved.

The steps are:

(1) The verification of the proposed formula or theorem for some positive integral value of n, usually the smallest. (Of course, we would not attempt to prove an unknown theorem by mathematical induction without first verifying it for several values of n.)

(2) The proof that if the proposed formula or theorem is true for $n = k$, some positive integer, it is true also for $n = k + 1$.

(3) The conclusion that the proposed formula or theorem is true for all values of n greater than the one for which verification was made in Step 1.

Example 3-8. Prove $1 + 3 + 5 + \ldots + (2n - 1) = n^2$.

(1) The formula is true for $n = 1$ since $1 = 1^2$.

(2) Let us assume the formula true for $n = k$, any positive integer, that is, let us assume that

$$1 + 3 + 5 + \ldots + (2k - 1) = k^2 \qquad (3\text{-}1)$$

We wish to show that, when (3-1) is true, the proposed formula is then true for $n = k + 1$; that is, that

$$1 + 3 + 5 + \ldots + (2k - 1) + (2k + 1) = (k + 1)^2 \qquad (3\text{-}2)$$

 **Note!**

Statements (3-1) and (3-2) are obtained by replacing n in the proposed formula by k and $k + 1$, respectively. Now it is clear that the left member of (3-2) can be obtained from the left member of (3-1) by adding $(2k + 1)$. At this point, the proposed formula is true or false according as we do or do not obtain the right member of (3-2) when $(2k + 1)$ is added to the right member of (3-1).]

Adding $(2k + 1)$ to both members of (3-1), we have

$$1 + 3 + 5 + \ldots + (2k - 1) + (2k + 1) \quad = k^2 + (2k + 1)$$
$$= (k + 1)^2 \qquad (3\text{-}3)$$

Now (3-3) is identical with (3-2); thus, if the proposed formula is true for any positive integer $n = k$, it is true for the next positive integer $n = k + 1$.

(3) Since the formula is true for $n = k = 1$ (Step 1), it is true for $n = k + 1 = 2$; being true for $n = k = 2$, it is true for $n = k + 1 = 3$, and so on. Hence the formula is true for all positive integral values of n.

Solved Problem 3-8. Prove by mathematical induction that $x^{2n} - y^{2n}$ is divisible by $x + y$.

Solution.

(1) The theorem is true for $n = 1$, since $x^2 - y^2 = (x - y)(x + y)$ is divisible by $x + y$.

(2) Let us assume the theorem true for $n = k$, a positive integer; that is, let us assume $x^{2k} - y^{2k}$ is divisible by $x + y$. We wish to show that, when this is true, $x^{2k+2} - y^{2k+2}$ is divisible by $x + y$. Now,

$$x^{2k+2} - y^{2k+2} = \left(x^{2k+2} - x^2 y^{2k}\right) + \left(x^2 y^{2k} - y^{2k+2}\right) =$$
$$x^2 \left(x^{2k} - y^{2k}\right) + y^{2k} \left(x^2 - y^2\right)$$

In the first term, $(x^{2k} - y^{2k})$ is divisible by $(x + y)$ by assumption, and in the second term $(x^2 - y^2)$ is divisible by $(x + y)$ by Step 1; hence if the theorem is true for $n = k$, a positive integer, it is true for the next one, $n = k + 1$.

(3) Since the theorem is true for $n = k = 1$, it is true for $n = k + 1 = 2$; being true for $n = k = 2$, it is true for $n = k + 1 = 3$; and so on, for every positive integral value of n.

The Binomial Theorem

By actual multiplication,

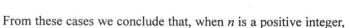

$$(a+b)^1 = a+b$$
$$(a+b)^2 = a^2 + 2ab + b^2$$
$$(a+b)^3 = a^3 + 3a^2 b + 3ab^2 + b^3$$
$$(a+b)^4 = a^4 + 4a^3 b + 6a^2 b^2 + 4ab^3 + b^4$$
$$(a+b)^5 = a^5 + 5a^4 b + 10a^3 b^2 + 10a^2 b^3 + 5ab^4 + b^5 \text{, etc.}$$

From these cases we conclude that, when n is a positive integer,

$$(a+b)^n = a^n + na^{n-1}b + \frac{n(n-1)}{1 \cdot 2}a^{n-2}b^2 + \frac{n(n-1)(n-2)}{1 \cdot 2 \cdot 3}a^{n-3}b^3$$
$$+ \cdots + nab^{n-1} + b^n$$

and note the following properties:

(1) The number of terms in the expansion is $(n + 1)$.

(2) The first term a of the binomial enters the first term of the expansion with exponent n, the second term with exponent $(n - 1)$, the third term with exponent $(n - 2)$, and so on.

(3) The second term b of the binomial enters the second term of the expansion with exponent 1, the third term with exponent 2, the fourth term with exponent 3, and so on.

(4) The sum of the exponents of a and b in any term is n.

(5) The coefficient of the first term in the expansion is 1, of the second term is $n/1$, of the third term is $\dfrac{n(n-1)}{1\cdot2}$, of the fourth term is $\dfrac{n(n-1)(n-2)}{1\cdot2\cdot3}$, etc.

(6) The coefficients of terms equidistant from the ends of the expansion are the same. Note that the number of factors in the numerator and denominator of any coefficient except the first and last is then either the exponent of a or of b, whichever is the smaller.

The above properties may be proved by mathematical induction.

Solved Problem 3-9. Expand $\left(3x+2y^2\right)^5$ and simplify term by term.

Solution. We put the several powers of $(3x)$ in first, then the powers of $(2y^2)$, and finally the coefficients, recalling Property 6.

$$\left(3x+2y^2\right)^5 = (3x)^5 + \frac{5}{1}(3x)^4\left(2y\right)^2 + \frac{5\cdot4}{1\cdot2}(3x)^3\left(2y^2\right)^2 +$$
$$\frac{5\cdot4}{1\cdot2}(3x)^2\left(2y^2\right)^3 + \frac{5}{1}(3x)\left(2y^2\right)^4 + \left(2y^2\right)^5$$

$$= 3^5x^5 + 5\cdot3^4x^4\cdot2y^2 + 10\cdot3^3x^3\cdot2^2y^4 + 10\cdot3^2x^2\cdot2^3y^6 +$$
$$5\cdot3x\cdot2^4y^8 + 2^5y^{10}$$

$$= 243x^5 + 810x^4y^2 + 1080x^3y^4 + 720x^2y^6 + 240xy^8 + 32y^{10}$$

The rth term $(r \le n + 1)$ in the expansion of $(a + b)^n$ is

$$\frac{n(n-1)(n-2)\cdots(n-r+2)}{1\cdot 2\cdot 3\cdots(r-1)}a^{n-r+1}b^{r-1}$$

Solved Problem 3-10. Find the seventh term of $(a + b)^{15}$ and simplify.

Solution. In the seventh term, the exponent of b is $7 - 1 = 6$, the exponent of a is $15 - 6 = 9$, and the coefficient has six factors in the numerator and denominator. Hence the term is

$$\frac{15\cdot 14\cdot 13\cdot 12\cdot 11\cdot 10}{1\cdot 2\cdot 3\cdot 4\cdot 5\cdot 6}a^9 b^6 = 5005a^9 b^6$$

When the laws above are used to expand $(a + b)^n$, where n is real but not a positive integer, an endless succession of terms is obtained. Such expansions are valid when $|b| < |a|$.

Solved Problem 3-11. Write the first five terms in the expansion of $(a+b)^{-3}$, $|b| < |a|$.

Solution

$$(a+b)^{-3} = a^{-3} + (-3)a^{-4}b + \frac{(-3)(-4)}{1\cdot 2}a^{-5}b^2 + \frac{(-3)(-4)(-5)}{1\cdot 2\cdot 3}a^{-6}b^3$$
$$+ \frac{(-3)(-4)(-5)(-6)}{1\cdot 2\cdot 3\cdot 4}a^{-7}b^4 + \cdots$$

$$= \frac{1}{a^3} - \frac{3b}{a^4} + \frac{6b^2}{a^5} - \frac{10b^3}{a^6} + \frac{15b^4}{a^7} - \cdots$$

Infinite Sequences

General Term of a Sequence

Frequently, the law of formation of a given sequence may be stated by giving a representative or *general term* of the sequence. This general term is a function of n, where n is the number of the term in the sequence. For this reason, it is called the nth term of the sequence.

Solved Problem 3-12. Write the first four terms of the sequence whose general term is: (a) $1/n$ and (b) $(-1)^{n-1} \dfrac{2n}{n^2+1}$.

Solution. (a) The first term ($n = 1$) is $\dfrac{1}{1} = 1$, the second term is ($n = 2$) is $\dfrac{1}{2}$, and so on. The first four terms are $1, \dfrac{1}{2}, \dfrac{1}{3}, \dfrac{1}{4}$.

(b) The first term ($n = 1$) is $(-1)^{1-1} \dfrac{2 \cdot 1}{1^2+1} = 1$, the second term ($n = 2$) is $(-1)^1 \dfrac{2 \cdot 2}{2^2+1} = -\dfrac{4}{5}$, and so on. The first four terms are

$$1, -\frac{4}{5}, \frac{3}{5}, -\frac{8}{17}.$$

Note that the effect of the factor $(-1)^{n-1}$ is to produce a sequence whose terms have alternate signs, the sign of the first term being positive. The same pattern of signs is also produced by the factor $(-1)^{n+1}$. In order to produce a sequence whose terms alternate in sign, the first term being negative, the factor $(-1)^n$ is used.

Solved Problem 3-13. Obtain the general term for each of the sequences: (a) 1, 4, 9, 16, 25, ... and (b) 3, 7, 11, 15, 19, 23, ...

Solution. (a) The terms of the sequence are the squares of the positive integers; the general term is n^2. (b) This is an arithmetic progression having $a = 3$ and $d = 4$. The general term is $a + (n - 1)d = 4n - 1$. Note, however, that the general term can be obtained about as easily by inspection.

Limit of an Infinite Sequence

If, for an infinite sequence

$$s_1, s_2, s_3, ..., s_n, ... \tag{3-4}$$

and a positive number ε, however small, there exists a number s and a positive integer m such that for all $n > m$

$$|s - s_n| < \varepsilon$$

then the limit of the sequence is s.

The statement that the limit of the sequence (3-4) is s describes the behavior of s^n as n increases without bound over the positive integers. Since we shall repeatedly be using the phrase "as n increases without bound" or the phrase "as n becomes infinite," which we shall take to be equivalent to the former phrase, we shall introduce the notation $n \to \infty$ for it. Thus, the behavior of s^n may be described briefly by

$$\lim_{n \to \infty} s_n = s$$

(read: the limit of s^n, as n becomes infinite, is s).

We state, without proof, the following theorem:

If $\lim\limits_{n\to\infty} s_n = s$ and $\lim\limits_{n\to\infty} t_n = t$, then

(a) $\lim\limits_{n\to\infty} \left(s_n \pm t_n \right) = \lim\limits_{n\to\infty} s_n \pm \lim\limits_{n\to\infty} t_n = s \pm t$

(b) $\lim\limits_{n\to\infty} \left(s_n \cdot t_n \right) = \lim\limits_{n\to\infty} s_n \cdot \lim\limits_{n\to\infty} t_n = s \cdot t$

(c) $\lim\limits_{n\to\infty} \dfrac{s_n}{t_n} = \dfrac{\lim\limits_{n\to\infty} s_n}{\lim\limits_{n\to\infty} t_n} = \dfrac{s}{t}$, provided $t \neq 0$

or, in words, if each of two variables approaches a limit, then the limits of the sum, difference, product, and quotient of the two variables are equal, respectively, to the sum, difference, product, and quotient of their limits provided only that, in the case of the quotient, the limit of the denominator is not zero.

This theorem makes it possible to find the limit of a sequence directly from its general term. In this connection, we shall need

$\lim\limits_{n\to\infty} a = a$, where a is any constant

$\lim\limits_{n\to\infty} \dfrac{1}{n^k} = 0$, $k > 0$

$\lim\limits_{n\to\infty} \dfrac{1}{b^n} = 0$, where b is a constant > 1

The following theorems are useful in establishing whether or not certain sequences have a limit.

1. Suppose M is a fixed number, such that, for all values of n,

$$s_n \leq s_{n+1} \quad \text{and} \quad s_n \leq M$$

then $\lim_{n\to\infty} s_n$ exists and is $\leq M$.

If, however, s_n eventually exceeds M, no matter how large M may be, $\lim_{n\to\infty} s_n$ does not exist.

2. Suppose M is a fixed number, such that, for all values of n,

$$s_n \geq s_{n+1} \quad \text{and} \quad s_n \geq M$$

then $\lim_{n\to\infty} s_n$ exists and is $\geq M$.

If, however, s_n is eventually smaller than M, no matter how small M may be, $\lim_{n\to\infty} s_n$ does not exist.

Recursively Defined Sequences

Sequences can be defined recursively. For example, suppose that $a_1 = 1$ and $a_{n+1} = 2a_n$ for every natural number n. Then,

$$a_1 = 1, \qquad a_2 = a_{1+1} = 2a_1 = 2, \qquad a_3 = a_{2+1} = 2a_2 = 4, \text{ etc.}$$

Thus, the sequence is 1, 2, 4, 8, ...

Infinite Series

The indicated sum of the terms of an infinite sequence is called an *infinite series*. Let

$$s_1 + s_2 + s_3 + \cdots + s_n + \cdots \tag{3-5}$$

be such a series and define the sequence of *partial sums*

$$S_1 = s_1, \quad S_2 = s_1 + s_2, \quad ..., \quad S_n = s_1 + s_2 + \cdots + s_n.$$

If $\lim_{n\to\infty} S_n$ exists, the series (3-5) is called *convergent*; if $\lim_{n\to\infty} S_n = S$, the series is said to converge to S. If $\lim_{n\to\infty} S_n$ does not exist, the series is called *divergent*.

Example 3-9. (a) Every infinite geometric series

$$a + ar + ar^2 + ... + ar^{n-1} + ...$$

is convergent if $|r| < 1$ and is divergent if $|r| \geq 1$.

(b) The *harmonic series* $1 + \dfrac{1}{2} + \dfrac{1}{3} + \cdots + \dfrac{1}{n} + \cdots$ is divergent.

A necessary condition that (3-5) be convergent is $\lim_{n\to\infty} s_n = 0$; that is, if (3-5) is convergent, then $\lim_{n\to\infty} s_n = 0$. However, this condition is *not* *sufficient* since the harmonic series is divergent although

$$\lim_{n\to\infty} s_n = \lim_{n\to\infty} \left(\frac{1}{n}\right) = 0.$$

A sufficient condition that (3-5) be divergent is $\lim_{n\to\infty} s_n \neq 0$; that is, if $\lim_{n\to\infty} s_n$ exists and is different from 0, or if $\lim_{n\to\infty} s_n$ does not exist, the series is divergent. This, in turn, is not a necessary condition since the harmonic series is divergent although $\lim\limits_{n\to\infty} s_n = 0$.

Series of Positive Terms

Comparison Test for Convergence of a Series of Positive Terms

1. If every term of a given series of positive terms is less than or equal to the corresponding term of a known convergent series from some point on in the series, the given series is convergent.

2. If every term of a given series of positive terms is equal to or greater than the corresponding term of a known divergent series from some point on in the series, the given series is divergent.

The following series will be found useful in making comparison tests:

(a) The geometric series $a + ar + ar^2 + \ldots + ar^n + \ldots$ which converges when $|r| < 1$ and diverges when $|r| \geq 1$.

(b) The p series $1 + \dfrac{1}{2^p} + \dfrac{1}{3^p} + \cdots + \dfrac{1}{n^p} + \cdots$ which converges for $p > 1$ and diverges for $p \leq 1$.

(c) Each new series tested.

In comparing two series, it is not sufficient to examine the first few terms of each series. *The general terms must be compared*, since the comparison must be shown from some point on.

The Ratio Test for Convergence

If, in a series of positive terms, the *test ratio*

$$r_n = \frac{s_{n+1}}{s_n}$$

approaches a limit R and $n \to \infty$, the series is convergent if $R < 1$ and is divergent if $R > 1$. If $R = 1$, the test fails to indicate convergency or divergency.

Series with Negative Terms

A series with all its terms negative may be treated as the negative of a series with all of its terms positive.

Alternating Series

A series whose terms are alternately positive and negative, as

$$s_1 - s_2 + s_3 - \cdots + (-1)^{n-1} s_n + \cdots \tag{3-6}$$

where each s is positive, is called an *alternating series*.

An alternating series (3-6) is convergent provided $s_n \geq s_{n+1}$, for every value of n, and $\lim\limits_{n \to \infty} s_n = 0$.

Absolutely Convergent Series

A series (3-5) $s_1 + s_2 + s_3 + \cdots + s_n + \cdots$ in which some of the terms are positive and some are negative is called *absolutely convergent* if the series of absolute values of the terms

$$|s_1| + |s_2| + |s_3| + \cdots + |s_n| + \cdots$$

is convergent.

Conditionally Convergent Series

A series (3-5), where some of the terms are positive and some are negative, is called *conditionally convergent* if it is convergent but the series of absolute values of its terms is divergent.

Example 3-10. The series $1 - \dfrac{1}{2} + \dfrac{1}{3} - \dfrac{1}{4} + \cdots$ is convergent, but the

series of absolute values of its terms $1 + \dfrac{1}{2} + \dfrac{1}{3} + \dfrac{1}{4} + \cdots$ is divergent.

Thus, the given series is conditionally convergent.

The Generalized Ratio Test

Let (3-5) be a series some of whose terms are positive and some are negative. Let

$$\lim_{n \to \infty} \frac{|s_{n+1}|}{|s_n|} = R$$

The series (3-5) is absolutely convergent if $R < 1$ and is divergent if $R > 1$. If $R = 1$, the test fails.

Power Series

Infinite series of the form

$$c_o + c_1 x + c_2 x^2 + \cdots + c_{n-1} x^{n-1} + \cdots \tag{3-7}$$

and

$$c_o + c_1 (x-a) + c_2 (x-a)^2 + \cdots + c_{n-1} (x-a)^{n-1} + \cdots \tag{3-8}$$

where a, c_o, c_1, c_a... are constants, are called *power series*. The first is called a power series in x and the second a power series in $(x - a)$.

The power series (3-7) converges for $x = 0$ and (3-8) converges for $x = a$. Both series may converge for other values of x but not necessarily for every finite value of x. Our problem is to find for a given power series all values of x for which the series converges. In finding this set of values, called the *interval of convergence* of the series, the generalized ratio test will be used.

Chapter 4
PERMUTATIONS, COMBINATIONS, AND PROBABILITY

IN THIS CHAPTER:

✔ *Permutations*
✔ *Combinations*
✔ *Probability*

Permutations

Any arrangement of a set of objects in a definite order is called a *permutation* of the set taken all at a time. For example, *abcd, acbd, bdca* are permutations of the set of letters *a*, *b*, *c*, *d* taken all at a time.

If a set contains *n* objects, any ordered arrangement of any $r \leq n$ of the objects is called a permutation of the *n* objects taken *r* at a time. For example, *ab, ba, ca, db* are permutations of the *n* = 4 letters *a*, *b*, *c*, *d* taken *r* = 2 at a time, while *abc, adb, bad, cad* are permutations of the *n* = 4 letters taken *r* = 3 at a time.

79

The number of permutations which may be formed in each situation can be found by means of the *fundamental principle*:

If one thing can be done in u different ways, if after it has been done in any one of these a second thing can be done in v different ways, if after it has been done in any one of these a third thing can be done in w different ways, . . . , then several things can be done in the order stated in $u \cdot v \cdot w$ ... different ways.

Solved Problem 4-1. In how many ways can 6 students be assigned to (a) a row of 6 seats, (b) a row of 8 seats?

Solution. (a) Let the seats be denoted xxxxxx. The seat on the left may be assigned to any one of the 6 students, that is, it may be assigned in 6 different ways. After the assignment has been made, the next seat may be assigned to any one of the 5 remaining students. After the assignment has been made, the next seat may be assigned to any one of the 4 remaining students, and so on. Placing the number of ways in which each seat may be assigned under the x marking the seat, we have

x	x	x	x	x	x
6	5	4	3	2	1

By the fundamental principle, the seats may be assigned in

$$6 \cdot 5 \cdot 4 \cdot 3 \cdot 2 \cdot 1 = 720 \text{ ways}$$

The reader should be assured that the seats might have been assigned to the students with the same result.

(b) Here each student must be assigned a seat. The first student may be assigned any one of the 8 seats, the second student any one of the 7 remaining seats, and so on. Letting x represent a student, we have

x	x	x	x	x	x
8	7	6	5	4	3

and the assignment may be made in $8 \cdot 7 \cdot 6 \cdot 5 \cdot 4 \cdot 3 = 20\ 160$ ways.

Solved Problem 4-2. Using the letters of the word MARKING and calling any arrangement a word, (a) how many different 7-letter words can be formed, (b) how many different 3-letter words can be formed?

Solution. (a) We must fill each of the positions xxxxxxx with a different letter. The first position may be filled in 7 ways, the second in 6 ways, and so on. Thus, we have

x	x	x	x	x	x	x
7	6	5	4	3	2	1

and there are $7 \cdot 6 \cdot 5 \cdot 4 \cdot 3 \cdot 2 \cdot 1 = 5040$ words.

(b) We must fill each of the positions xxx with a different letter. The first position can be filled in 7 ways, the second in 6 ways, and the third in 5 ways. Thus, there are $7 \cdot 6 \cdot 5 = 210$ words.

Define $n!$ (n factorial) to be

$$n \cdot (n-1) \cdot (n-2) \cdots (2)(1)$$

(where $0! = 1! = 1$). Then, if we define $\binom{n}{k}$ to be $\dfrac{n!}{k!(n-k)!}$, we call $\binom{n}{k}$ a binomial coefficient and note that

$$(a+b)^n = \sum_{k=0}^{n} \binom{n}{k} a^{n-k} b^k , \qquad n \geq 1$$

Permutations of Objects Not All Different

If there are n objects of which k are alike while the remaining $(n - k)$ objects are different from them and from each other, it is clear that the number of different permutations of the n objects taken all together is not $_nP_n$.

Solved Problem 4-3. How many different permutations of four letters can be formed using the letters of the word *bass*?

Solution. For the moment, think of the letters as b, a, s_1, s_2 so that they are all different. Then

$$bas_1s_2 \quad as_1bs_2 \quad s_2s_1ba \quad s_1as_2b \quad bs_1s_2a$$

$$bas_2s_1 \quad as_2bs_1 \quad s_1s_2ba \quad s_2as_1b \quad bs_2s_1a$$

are 10 of the 24 permutations of the four letters taken all together. However, when the subscripts are removed, it is seen that the two permutations in each column are alike.

Thus, there are $\dfrac{1 \cdot 2 \cdot 3 \cdot 4}{1 \cdot 2} = 12$ different permutations.

In general, given n objects of which k_1 are of one sort, k_2 of another, k_3 of another, . . .; then the number of different permutations that can be made from the n objects taken all together is

$$\frac{n!}{k_1!k_2!k_3!...}$$

Combinations

The combinations of n objects taken r at a time consist of all possible sets of r of the objects, without regard to the order of arrangement. The number of combinations of n objects taken r at a time will be denoted by $_nC_r$.

For example, the combinations of the $n = 4$ letters a, b, c, d taken $r = 3$ at a time, are

abc, abd, acd, bcd

Thus, $_4C_3 = 4$. When the letters of each combination are rearranged (in 3! ways), we obtain the $_4P_3$ permutations of the 4 letters taken 3 at a time. Hence, $_4P_3 = 3!(_4C_3)$ and $_4C_3 = \dfrac{_4P_3}{3!}$

The number of combinations of n different objects taken r at a time is equal to the number of permutations of the n objects taken r at a time divided by factorial r, or

$$_nC_r = \frac{_nP_r}{r!} = \frac{n(n-1)\cdots(n-r+1)}{1\cdot 2\cdots r}$$

Solved Problem 4-4. From a shelf containing 12 different toys, a child is permitted to select 3. In how many ways can this be done?

Solution. The required number is

$$_{12}C_3 = \frac{_{12}P_3}{3!} = \frac{12\cdot 11\cdot 10}{1\cdot 2\cdot 3} = 220$$

Notice that $_nC_r$ is the rth term's coefficient in the binomial theorem.

Solved Problem 4-5. A committee of 5 is to be selected from 12 seniors and 8 juniors. In how many ways can this be done (a) if the committee is to consist of 3 seniors and 2 juniors, (b) if the committee is to contain at least 3 seniors and 1 junior?

Solution. (a) With each of the $_{12}C_3$ selections of 3 seniors, we may associate any one of the $_8C_2$ selections of 2 juniors. Thus, a committee can be selected in $_{12}C_3 \cdot _8C_2 = \dfrac{12 \cdot 11 \cdot 10}{1 \cdot 2 \cdot 3} \cdot \dfrac{8 \cdot 7}{1 \cdot 2} = 6160$ ways.

(b) The committee may consist of 3 seniors and 2 juniors or of 4 seniors and 1 junior. A committee of 3 seniors and 2 juniors can be selected in 6160 ways, and a committee of 4 seniors and 1 junior can be selected in $_{12}C_4 \cdot _8C_1 = 3960$ ways. In all, a committee may be selected in 6160 + 3960 = 10120 ways.

Solved Problem 4-6. There are ten points A, B, …, in a plane on the same straight line. (a) How many lines are determined by the points? (b) How many of the lines pass through A? (c) How many triangles are determined by the points? (d) How many of the triangles have A as a vertex? (e) How many of the triangles have AB as a side?

Solution

(a) Since any two points determine a line, there are $_{10}C_2 = 45$ lines .

(b) To determine a line through A, one other point must be selected. Thus, there are nine lines through A.

(c) Since any three of the points determine a triangle, there are $_{10}C_3 = 120$ triangles.

(d) Two additional points are needed to form a triangle. These points may be selected from the nine points in $_9C_2 = 36$ ways .

(e) One additional point is needed; there are eight triangles having AB as a side.

Solved Problem 4-7. The English alphabet consists of 21 consonants and 5 vowels. (a) In how many ways can 4 consonants and 2 vowels be selected? (b) How many words consisting of 4 consonants and 2 vowels can be formed? (c) How many of the words in (b) begin with R? (d) How many of the words in (c) contain E?

Solution

(a) The 4 consonants can be selected in $_{21}C_4$ ways and the 2 vowels can be selected in $_5C_2$ ways. Thus, the selections may be made in $_{21}C_4 \cdot _5C_2 = 59\ 850$ ways.

(b) From each of the selections in (a), 6! words may be formed by permuting the letters. Therefore, $59850 \cdot 6! = 43\ 092\ 000$ words can be formed.

(c) Since the position of the consonant R is fixed, we must select 3 other consonants (in $_{20}C_3$ ways) and 2 vowels (in $_5C_2$ ways), and arrange each selection of 5 letters in all possible ways. Thus, there are $_{20}C_3 \cdot _5C_2 \cdot 5! = 1\ 368\ 000$ words.

(d) Since the position of the consonant R is fixed but the position of the vowel E is not, we must select 3 other consonants (in $_{20}C_3$ ways) and 1 other vowel (in four ways), and arrange each set of 5 letters in all possible ways. Thus, there are $_{20}C_3 \cdot 4 \cdot 5! = 547\ 200$ words.

Probability

In estimating the probability that a given event will or will not happen, we may, as in the case of drawing a face card from an ordinary deck, count the different ways in which this event may or may not happen. On the other hand, in the case of estimating the probability that a person who is now 25 years old will live to receive a bequest at age 30, we are forced to depend upon such knowledge of what has happened on similar occasions in the past as is available. In the first case, the result is called *mathematical* or *theoretical probability*; in the latter case, the result is called *statistical* or *empirical probability*.

Mathematical Probability

If an event must result in some one of n, $(n \neq 0)$ different but *equally likely ways* and if a certain s of these ways are considered successes and the other $f = n - s$ ways are considered failures, then the probability of success in a given trial is $p = \dfrac{s}{n}$ and the probability of failure is $q = \dfrac{f}{n}$.

Since $p + q = \dfrac{s+f}{n} = \dfrac{n}{n} = 1$, $p = 1 - q$ and $q = 1 - p$.

Solved Problem 4-8. One card is drawn from an ordinary deck. What is the probability (a) that it is a red card, (b) that it is a spade, (c) that it is a king, (d) that it is not the ace of hearts?

Solution. One card is drawn from the deck in $n = 52$ different ways.

(a) A red card can be drawn from the deck in $s = 26$ different ways. Thus, the probability of drawing a red card is $\dfrac{s}{n} = \dfrac{26}{52} = \dfrac{1}{2}$.

(b) A spade can be drawn from the deck in 13 different ways. The probability of drawing a spade is $\dfrac{13}{52} = \dfrac{1}{4}$.

(c) A king can be drawn in 4 ways. The required probability is $\dfrac{4}{52} = \dfrac{1}{13}$.

(d) The ace of hearts can be drawn in 1 way; the probability of drawing the ace of hearts is $\dfrac{1}{52}$. Thus the probability of *not* drawing the ace of hearts is $1 - \dfrac{1}{52} = \dfrac{51}{52}$.

Two or more events are called *mutually exclusive* if not more than one of them can occur in a single trial. Thus, the drawing of a jack and the drawing of a queen on a single draw from an ordinary deck are mutually exclusive events; however, the drawing of a jack and the drawing of a spade are not mutually exclusive.

Solved Problem 4-9. Find the probability of drawing a jack or a queen from an ordinary deck of cards.

Solution. Since there are four jacks and four queens, $s = 8$ and $p = \dfrac{8}{52} = \dfrac{2}{13}$. Now the probability of drawing a jack is $\dfrac{1}{13}$, the probability of drawing a queen is $\dfrac{1}{13}$, and the required probability is $\dfrac{1}{13} + \dfrac{1}{13} = \dfrac{2}{13}$.

We write $P(A)$ to denote the "probability that A occurs." We have verified:

THEOREM A. The probability that some one of a set of mutually exclusive events will happen at a single trial is the sum of their separate probabilities of happening. The probability that A will occur or B will occur,

$$P(A \cup B) = P(A) + P(B) - P(A \cap B)$$
.

Two events A and B are called *independent* if the happening of one does not affect the happening of the other. Thus, in a toss of two dice, the fall of either does not affect the fall of the other. However, in drawing two cards from a deck, the probability of obtaining a red card on the second draw depends upon whether or not a red card was obtained on the draw of the first card. Two such events are called *dependent*. More explicitly, if $P(A \cap B) = P(A) \cdot P(B)$, then A and B are independent.

Solved Problem 4-10. One bag contains 4 white and 4 black balls, a second bag contains 3 white and 6 black balls, and a third contains 1 white and 5 black balls. If one ball is drawn from each bag, find the probability that all are white.

Solution. A ball can be drawn from the first bag in any one of 8 ways, from the second in any one of 9 ways, and from the third in any one of 6 ways; hence three balls can be drawn one from each bag in $8 \cdot 9 \cdot 6$ ways. A white ball can be drawn from the first bag in 4 ways, from the second in 3 ways, and from the third in 1 way; hence three white balls can be drawn one from each bag in $4 \cdot 3 \cdot 1$ ways. Thus the required probability is

$$\frac{4 \cdot 3 \cdot 1}{8 \cdot 9 \cdot 6} = \frac{1}{36}$$

Now drawing a white ball from one bag does not affect the drawing of a white ball from another so that here we are concerned with three independent events. The probability of drawing a white ball from the first bag is $\frac{4}{8}$, from the second is $\frac{3}{9}$, and from the third bag is $\frac{1}{6}$.

Since the probability of drawing three white balls, one from each bag, is $\frac{4}{8} \cdot \frac{3}{9} \cdot \frac{1}{6}$, we have verified

THEOREM B. The probability that all of a set of independent events will happen in a single trial is the product of their separate probabilities.

THEOREM C (concerning dependent events). If the probability that an event will happen is p_1, and if after it has happened, the probability that a second event will happen is p_2, the probability that the two events will happen in that order is $p_1 p_2$.

Solved Problem 4-11. Two cards are drawn from an ordinary deck. Find the probability that both are face cards (king, queen, jack) if (a) the first card drawn is replaced before the second is drawn, (b) the first card drawn is not replaced before the second is drawn.

Solution. (a) Since each drawing is made from a complete deck, we have the case of two independent events. The probability of drawing a face card in a single draw is $\dfrac{12}{52}$; thus, the probability of drawing two face cards, under the conditions imposed, is $\dfrac{12}{52} \cdot \dfrac{12}{52} = \dfrac{9}{169}$.

(b) Here the two events are dependent. The probability that the first drawing results in a face card is $\dfrac{12}{52}$. Now, of the 51 cards remaining in the deck, there are 11 face cards; the probability that the second drawing results in a face card is $\dfrac{11}{51}$. Hence the probability of drawing two face cards is $\dfrac{12}{52} \cdot \dfrac{11}{51} = \dfrac{11}{221}$.

Solved Problem 4-12. Two dice are tossed six times. Find the probability (a) that 7 will show on the first four tosses and will not show on the other two, (b) that 7 will show on exactly four of the tosses.

The probability that 7 will show on a single toss is $p = \dfrac{1}{6}$ and the probability that 7 will not show is $q = 1 - p = \dfrac{5}{6}$.

Solution. (a) The probability that 7 will show on the first four tosses and will not show on the other two is

$$\frac{1}{6} \cdot \frac{1}{6} \cdot \frac{1}{6} \cdot \frac{1}{6} \cdot \frac{5}{6} \cdot \frac{5}{6} = \frac{25}{46\ 656}$$

(b) The four tosses on which 7 is to show may be selected in $_6C_4 = 15$ ways. Since these 15 ways constitute mutually exclusive events and the probability of any one of them is $\left(\dfrac{1}{6}\right)^4 \left(\dfrac{5}{6}\right)^2$, the probability that 7 will show exactly four times in six tosses is

$$_6C_4 \left(\frac{1}{6}\right)^4 \left(\frac{5}{6}\right)^2 = \frac{125}{15\ 552}\ .$$

We have verified

THEOREM D. If p is the probability that an event will happen and q is the probability that it will fail to happen at a given trial, the probability that it will happen exactly r times in n trials is $_nC_r\, p^r q^{n-r}$.

Empirical Probability

If an event has been observed to happen s times in n trials, the ratio

$p = \dfrac{s}{n}$ is defined as the *empirical probability* that the event will happen at any future trial. The confidence which can be placed in such probabilities depends in a large measure on the number of observations used. Life insurance companies, for example, base their premium rate on empirical probabilities. For this purpose, they use a mortality table based on an enormous number of observations over the years.

The American Experience Table of Mortality begins with 100 000 persons, all of age 10 years and indicates the number of the group who die each year thereafter. In using this table, it will be assumed that the laws stated above for mathematical probability hold also for empirical probability.

Solved Problem 4-13. Find the probability that a person 20 years old (a) will die during the year, (b) will die during the next 10 years, (c) will reach age 75.

Solution. (a) Of the 100 000 persons alive at age 10 years, 92 637 are alive at age 20 years. Of these 92 637, a total of 723 will die during the year. The probability that a person 20 years of age will die during the

year is $\dfrac{723}{92\ 637} = 0.0078$.

(b) Of the 92 637 who reach age 20 years, 85 441 reach age 30 years; thus 92 637 − 85 441 = 7196 die during the 10-year period. The required

probability is $\dfrac{7196}{92\ 637} = 0.0777$.

(c) Of the 92 637 alive at age 20 years, 26 237 will reach age 75 years.

The required probability is $\dfrac{26\ 237}{92\ 637} = 0.2832$.

Chapter 5
SYSTEMS OF LINEAR EQUATIONS USING DETERMINANTS

IN THIS CHAPTER:

✔ *Determinants of Orders Two and Three*
✔ *Determinants of Order* n
✔ *Systems of Linear Equations*

Determinants of Orders Two and Three

Determinants of Order Two

The symbol $\begin{vmatrix} a_1 & b_1 \\ a_2 & b_2 \end{vmatrix}$, consisting of 2^2 numbers called *elements* arranged in two rows and two columns, is called a *determinant of order two*. The elements a_1 and b_2 are said to lie along the *principal diagonal*; the elements a_2 and b_1 are said to lie along the *secondary diagonal*.

The *value* of the determinant is obtained by forming the product of the elements along the principal diagonal and subtracting the product of the elements along the secondary diagonal; thus,

$$\begin{vmatrix} a_1 & b_1 \\ a_2 & b_2 \end{vmatrix} = a_1 b_2 - a_2 b_1$$

Solved Problem 5-1. Evaluate each of the following determinants:

(a) $\begin{vmatrix} 2 & 3 \\ 4 & 5 \end{vmatrix}$ (b) $\begin{vmatrix} 5 & -2 \\ 3 & 1 \end{vmatrix}$

Solution

(a) $\begin{vmatrix} 2 & 3 \\ 4 & 5 \end{vmatrix} = 2 \cdot 5 - 4 \cdot 3 = 10 - 12 = -2$

(b) $\begin{vmatrix} 5 & -2 \\ 3 & 1 \end{vmatrix} = 5 \cdot 1 - 3(-2) = 5 - (-6) = 11$

The solution of the consistent and independent equations

$$a_1 x + b_1 y = c_1$$
$$a_2 x + b_2 y = c_2 \tag{5-1}$$

may be expressed as quotients of determinants of order two:

$$x = \frac{c_1 b_2 - c_2 b_1}{a_1 b_2 - a_2 b_1} = \frac{\begin{vmatrix} c_1 & b_1 \\ c_2 & b_2 \end{vmatrix}}{\begin{vmatrix} a_1 & b_1 \\ a_2 & b_2 \end{vmatrix}}, \quad y = \frac{a_1 c_2 - a_2 c_1}{a_1 b_2 - a_2 b_1} = \frac{\begin{vmatrix} a_1 & c_1 \\ a_2 & c_2 \end{vmatrix}}{\begin{vmatrix} a_1 & b_1 \\ a_2 & b_2 \end{vmatrix}}$$

These equations are consistent and independent if and only if

$$\begin{vmatrix} a_1 & b_1 \\ a_2 & b_2 \end{vmatrix} \neq 0$$

Solved Problem 5-2. Using determinants, solve

$$y = 3x + 1$$
$$4x + 2y - 7 = 0$$

Solution. Arrange the equations in the form (5-1):

$$3x - y = -1$$
$$4x + 2y = 7$$

The solution requires the values of three determinants:

(a) The denominator, D, formed by writing the coefficients of x and y in order

$$D = \begin{vmatrix} 3 & -1 \\ 4 & 2 \end{vmatrix} = 3 \cdot 2 - 4(-1) = 6 + 4 = 10$$

(b) The numerator of x, N^x, formed from D by replacing the coefficients of x by the constant terms

$$N_x = \begin{vmatrix} -1 & -1 \\ 7 & 2 \end{vmatrix} = -1 \cdot 2 - 7(-1) = -2 + 7 = 5$$

(c) The numerator of y, N^y, formed from D by replacing the coefficients of y by the constant terms

$$N_y = \begin{vmatrix} 3 & -1 \\ 4 & 7 \end{vmatrix} = 3 \cdot 7 - 4(-1) = 21 + 4 = 25$$

Then, $x = \dfrac{N_x}{D} = \dfrac{5}{10} = \dfrac{1}{2}$ and $y = \dfrac{N_y}{D} = \dfrac{25}{10} = \dfrac{5}{2}$.

Determinants of Order Three

The symbol

$$\begin{vmatrix} a_1 & b_1 & c_1 \\ a_2 & b_2 & c_2 \\ a_3 & b_3 & c_3 \end{vmatrix}$$

consisting of 3^2 elements arranged in three rows and three columns, is called a *determinant of order three*. Its value is

$$a_1 b_2 c_3 + a_2 b_3 c_1 + a_3 b_1 c_2 - a_1 b_3 c_2 - a_2 b_1 c_3 - a_3 b_2 c_1$$

This may be written as

$$a_1 \left(b_2 c_3 - b_3 c_2 \right) - b_1 \left(a_2 c_3 - a_3 c_2 \right) + c_1 \left(a_2 b_3 - a_3 b_2 \right)$$

or

$$a_1 \begin{vmatrix} b_2 & c_2 \\ b_3 & c_3 \end{vmatrix} - b_1 \begin{vmatrix} a_2 & c_2 \\ a_3 & c_3 \end{vmatrix} + c_1 \begin{vmatrix} a_2 & b_2 \\ a_3 & b_3 \end{vmatrix} \qquad (5\text{-}2)$$

to involve three determinants of order two. Note that the elements which multiply the determinants of order two are the elements of the first row of the given determinant. In all, six such representations using the elements of each of the three rows and each of the three columns may be worked out.

The solution of the system of consistent and independent equations

$$a_1 x + b_1 y + c_1 z = d_1$$
$$a_2 x + b_2 y + c_2 z = d_2$$
$$a_3 x + b_3 y + c_3 z = d_3$$

in determinant form is given by

$$x = \frac{N_x}{D} = \frac{\begin{vmatrix} d_1 & b_1 & c_1 \\ d_2 & b_2 & c_2 \\ d_3 & b_3 & c_3 \end{vmatrix}}{\begin{vmatrix} a_1 & b_1 & c_1 \\ a_2 & b_2 & c_2 \\ a_3 & b_3 & c_3 \end{vmatrix}}, \; y = \frac{N_y}{D} = \frac{\begin{vmatrix} a_1 & d_1 & c_1 \\ a_2 & d_2 & c_2 \\ a_3 & d_3 & c_3 \end{vmatrix}}{D}, \; z = \frac{N_z}{D} = \frac{\begin{vmatrix} a_1 & b_1 & d_1 \\ a_2 & b_2 & d_2 \\ a_3 & b_3 & d_3 \end{vmatrix}}{D}$$

The determinant D is formed by writing the coefficients of x, y, z in order while the determinant appearing in the numerator for any unknown is obtained from D by replacing the column of coefficients of that unknown by the column of constants.

The system is consistent and independent if and only if $D \neq 0$.

Solved Problem 5-3. Using determinants, solve

$$x + 3y + 2z = -13$$
$$2x - 6y + 3z = 32$$
$$3x - 4y - z = 12$$

Solution. The solution requires the values of four determinants:

(a) The denominator,

$$D = \begin{vmatrix} 1 & 3 & 2 \\ 2 & -6 & 3 \\ 3 & -4 & -1 \end{vmatrix}$$
$$= 1(6+12) - 3(-2-9) + 2(-8+18)$$
$$= 18 + 33 + 20$$
$$= 71$$

(b) The numerator of x,

$$N_x = \begin{vmatrix} -13 & 3 & 2 \\ 32 & -6 & 3 \\ 12 & -4 & -1 \end{vmatrix}$$

$$= -13(6+12)-3(-32-36)+2(-128+72)$$

$$= -234+204-112$$

$$= -142$$

(c) The numerator of y,

$$N_y = \begin{vmatrix} 1 & -13 & 2 \\ 2 & 32 & 3 \\ 3 & 12 & -1 \end{vmatrix}$$

$$= 1(-32-36)-(-13)(-2-9)+2(24-96)$$

$$= -68-143-144$$

$$= -355$$

(d) the numerator of z,

$$N_z = \begin{vmatrix} 1 & 3 & -13 \\ 2 & -6 & 32 \\ 3 & -4 & 12 \end{vmatrix}$$

$$= 1(-72+128)-3(24-96)+(-13)(-8+18)$$

$$= 56+216-130$$

$$= 142$$

Then,

$$x = \frac{N_x}{D} = \frac{-142}{71} = -2, \quad y = \frac{N_y}{D} = \frac{-355}{71} = -5, \quad z = \frac{N_z}{D} = \frac{142}{71} = 2$$

Determinants of Order n

A determinant of order n consists of n^2 numbers called elements arranged in n rows and columns, and enclosed by two vertical lines. For example,

$$\Delta_1 = |a_1| \quad \Delta_2 = \begin{vmatrix} a_1 & b_1 \\ a_2 & b_2 \end{vmatrix} \quad \Delta_3 = \begin{vmatrix} a_1 & b_1 & c_1 \\ a_2 & b_2 & c_2 \\ a_3 & b_3 & c_3 \end{vmatrix} \quad \Delta_4 = \begin{vmatrix} a_1 & b_1 & c_1 & d_1 \\ a_2 & b_2 & c_2 & d_2 \\ a_3 & b_3 & c_3 & d_3 \\ a_4 & b_4 & c_4 & d_4 \end{vmatrix}$$

are determinants of orders one, two, three, and four, respectively. In this notation, the letters designate columns and the subscripts designate rows. Thus, all elements with basal letter c are in the third column and all elements with subscript 2 are in the second row.

The minor of a given element of a determinant is the determinant of the elements which remain after deleting the row and the column in which the given element stands. For example, the minor of a_1 in Δ_4 is

$$\begin{vmatrix} b_2 & c_2 & d_2 \\ b_3 & c_3 & d_3 \\ b_4 & c_4 & d_4 \end{vmatrix}$$

and the minor of b^3 is

$$\begin{vmatrix} a_1 & c_1 & d_1 \\ a_2 & c_2 & d_2 \\ a_4 & c_4 & d_4 \end{vmatrix}$$

Note that the minor of a given element contains no element having either the letter or the subscript of the given element.

The value of a determinant of order one is the single element of the determinant. A determinant of order $n > 1$ may be expressed as the sum of n products formed by multiplying each element of any chosen row (column) by its minor and prefixing a proper sign. The proper sign associated with each product is $(-1)^{i+j}$ where i is the number of the row and j is the number of the column in which the element stands. For example,

$$\Delta_3 = -a_2 \begin{vmatrix} b_1 & c_1 \\ b_3 & c_3 \end{vmatrix} + b_2 \begin{vmatrix} a_1 & c_1 \\ a_3 & c_3 \end{vmatrix} - c_2 \begin{vmatrix} a_1 & b_1 \\ a_3 & b_3 \end{vmatrix}$$

is the expansion of Δ_3 along the second row. The sign given to the first product is $-$ since a_2 stands in the second row and first column, and $(-1)^{2+1} = -1$. In all, there are six expansions of Δ_3 along its rows and columns yielding identical results when the minors are evaluated. There are eight expansions of Δ_4 along its rows and columns, of which,

$$\Delta_4 = +a_1 \begin{vmatrix} b_2 & c_2 & d_2 \\ b_3 & c_3 & d_3 \\ b_4 & c_4 & d_4 \end{vmatrix} - b_1 \begin{vmatrix} a_2 & c_2 & d_2 \\ a_3 & c_3 & d_3 \\ a_4 & c_4 & d_4 \end{vmatrix} + c_1 \begin{vmatrix} a_2 & b_2 & d_2 \\ a_3 & b_3 & d_3 \\ a_4 & b_4 & d_4 \end{vmatrix} - d_1 \begin{vmatrix} a_2 & b_2 & c_2 \\ a_3 & b_3 & c_3 \\ a_4 & b_4 & c_4 \end{vmatrix}$$

(along the first row)

$$\Delta_4 = +a_1 \begin{vmatrix} b_2 & c_2 & d_2 \\ b_3 & c_3 & d_3 \\ b_4 & c_4 & d_4 \end{vmatrix} - a_2 \begin{vmatrix} b_1 & c_1 & d_1 \\ b_3 & c_3 & d_3 \\ b_4 & c_4 & d_4 \end{vmatrix} + a_3 \begin{vmatrix} b_1 & c_1 & d_1 \\ b_2 & c_2 & d_2 \\ b_4 & c_4 & d_4 \end{vmatrix} - a_4 \begin{vmatrix} b_1 & c_1 & d_1 \\ b_2 & c_2 & d_2 \\ b_3 & c_3 & d_3 \end{vmatrix}$$

(along the first column)

$$\Delta_4 = -a_4 \begin{vmatrix} b_1 & c_1 & d_1 \\ b_2 & c_2 & d_2 \\ b_3 & c_3 & d_3 \end{vmatrix} + b_4 \begin{vmatrix} a_1 & c_1 & d_1 \\ a_2 & c_2 & d_2 \\ a_3 & c_3 & d_3 \end{vmatrix} - c_4 \begin{vmatrix} a_1 & b_1 & d_1 \\ a_2 & b_2 & d_2 \\ a_3 & b_3 & d_3 \end{vmatrix} + d_4 \begin{vmatrix} a_1 & b_1 & c_1 \\ a_2 & b_2 & c_2 \\ a_3 & b_3 & c_3 \end{vmatrix}$$

(along the fourth row)

$$\Delta_4 = -b_1 \begin{vmatrix} a_2 & c_2 & d_2 \\ a_3 & c_3 & d_3 \\ a_4 & c_4 & d_4 \end{vmatrix} + b_2 \begin{vmatrix} a_1 & c_1 & d_1 \\ a_3 & c_3 & d_3 \\ a_4 & c_4 & d_4 \end{vmatrix} - b_3 \begin{vmatrix} a_1 & c_1 & d_1 \\ a_2 & c_2 & d_2 \\ a_4 & c_4 & d_4 \end{vmatrix} + b_4 \begin{vmatrix} a_1 & c_1 & d_1 \\ a_2 & c_2 & d_2 \\ a_3 & c_3 & d_3 \end{vmatrix}$$

(along the second column)

are examples.

The cofactor of an element of a determinant is the minor of that element together with the sign associated with the product of that element and its minor in the expansion of the determinant. The cofactors of the elements $a_1, a_2, b_1, b_3, c_1, \ldots$ will be denoted by $A_1, A_2, B_1, B_3, C_1, \ldots$. Thus, the cofactor of c_1 in Δ_3 is $C_1 = + \begin{vmatrix} a_2 & b_2 \\ a_3 & b_3 \end{vmatrix}$ and the cofactor of b_3 is $B_3 = - \begin{vmatrix} a_1 & c_1 \\ a_2 & c_2 \end{vmatrix}$.

When cofactors are used, the expansion of Δ_4 given above takes the more compact form

$$\begin{aligned}
\Delta_4 &= a_1 A_1 + b_1 B_1 + c_1 C_1 + d_1 D_1 && \text{(along the first row)} \\
&= a_1 A_1 + a_2 A_2 + a_3 A_3 + a_4 A_4 && \text{(along the first column)} \\
&= a_4 A_4 + b_4 B_4 + c_4 C_4 + d_4 D_4 && \text{(along the fourth row)} \\
&= b_1 B_1 + b_2 B_2 + b_3 B_3 + b_4 B_4 && \text{(along the first row)}
\end{aligned}$$

Properties of Determinants

Subject always to our assumption of equivalent expansions of a determinant along any of its rows or columns, the following theorems may be proved by mathematical induction.

THEOREM 1. If two rows (or two columns) of a determinant are identical, the value of the determinant is zero. For example,

$$\begin{vmatrix} 2 & 3 & 2 \\ 3 & 1 & 3 \\ 1 & 4 & 1 \end{vmatrix} = 0$$

COROLLARY 1. If each of the elements of a row (or a column) of a determinant is multiplied by the cofactor of the corresponding element of another row (or a column), the sum of the products is zero.

THEOREM 2. If the elements of a row (or a column) of a determinant are multiplied by any number m, the determinant is multiplied by m. For example,

$$5\begin{vmatrix} 2 & 3 & 4 \\ 3 & -1 & 2 \\ 1 & 4 & -3 \end{vmatrix} = \begin{vmatrix} 10 & 3 & 4 \\ 15 & -1 & 2 \\ 5 & 4 & -3 \end{vmatrix} = \begin{vmatrix} 2 & 3 & 4 \\ 15 & -5 & 10 \\ 1 & 4 & -3 \end{vmatrix}$$

THEOREM 3. If each of the elements of a row (or a column) of a determinant is expressed as the sum of two or more numbers, the determinant may be written as the sum of two or more determinants. For example,

$$\begin{vmatrix} 2 & 5 & 4 \\ 4 & -2 & 3 \\ 1 & -4 & 3 \end{vmatrix} = \begin{vmatrix} -2+4 & 5 & 4 \\ 3+1 & -2 & 3 \\ 1+0 & -4 & 3 \end{vmatrix} = \begin{vmatrix} -2 & 5 & 4 \\ 3 & -2 & 3 \\ 1 & -4 & 3 \end{vmatrix} + \begin{vmatrix} 4 & 5 & 4 \\ 1 & -2 & 3 \\ 0 & -4 & 3 \end{vmatrix}$$

THEOREM 4. If, to the elements of any row (or any column) of a determinant, there is added m times the corresponding elements of another row (or another column), the value of the determinant is unchanged. For example,

$$\begin{vmatrix} -2 & 5 & 4 \\ 3 & -2 & 2 \\ 1 & -4 & 3 \end{vmatrix} = \begin{vmatrix} -2 & 5+4(-2) & 4 \\ 3 & -2+4(3) & 2 \\ 1 & -4+4(1) & 3 \end{vmatrix} = \begin{vmatrix} -2 & -3 & 4 \\ 3 & 10 & 2 \\ 1 & 0 & 3 \end{vmatrix}$$

Evaluation of Determinants

A determinant of any order may be evaluated by expanding it and all subsequent determinants (or minors) thus obtained along a row or column. This procedure may be greatly simplified by the use of Theorem 4.

Solved Problem 5-4. Evaluate

$$\begin{vmatrix} 1 & 4 & 3 & 1 \\ 2 & 8 & 2 & 5 \\ 4 & -4 & -1 & -3 \\ 2 & 5 & 3 & 3 \end{vmatrix}$$

Solution. Using the first column since it contains the element 1 in the first row, we obtain an equivalent determinant all of whose elements, except the first, in the first row are zeros. We have

$$\begin{vmatrix} 1 & 4 & 3 & 1 \\ 2 & 8 & 2 & 5 \\ 4 & -4 & -1 & -3 \\ 2 & 5 & 3 & 3 \end{vmatrix} = \begin{vmatrix} 1 & 4+(-4)1 & 3+(-3)1 & 1+(-1)1 \\ 2 & 8+(-4)2 & 2+(-3)2 & 5+(-1)2 \\ 4 & -4+(-4)4 & -1+(-3)4 & -3+(-1)4 \\ 2 & 5+(-4)2 & 3+(-3)2 & 3+(-1)2 \end{vmatrix}$$

$$= \begin{vmatrix} 1 & 0 & 0 & 0 \\ 2 & 0 & -4 & 3 \\ 4 & -20 & -13 & -7 \\ 2 & -3 & -3 & 1 \end{vmatrix}$$

$$= \begin{vmatrix} 0 & -4 & 3 \\ -20 & -13 & -7 \\ -3 & -3 & 1 \end{vmatrix}$$

(by expanding along the first row)

Expanding the resulting determinant along the first row to take full advantage of the element 0, we have

$$\begin{vmatrix} 0 & -4 & 3 \\ -20 & -13 & -7 \\ -3 & -3 & 1 \end{vmatrix} = 4(-20-21)+3(60-39)=-101$$

Systems of n Linear Equations

Systems of Linear Equations in n Unknowns

Consider, for the sake of brevity, the system of four linear equations in four unknowns

$$\begin{aligned} a_1 x + b_1 y + c_1 z + d_1 w &= k_1 \\ a_2 x + b_2 y + c_2 z + d_2 w &= k_2 \\ a_3 x + b_3 y + c_3 z + d_3 w &= k_3 \\ a_4 x + b_4 y + c_4 z + d_4 w &= k_4 \end{aligned} \qquad (5\text{-}3)$$

in which each equation is written with the unknowns x, y, z, w in that order on the left side and the constant term on the right side. Form

$$D = \begin{vmatrix} a_1 & b_1 & c_1 & d_1 \\ a_2 & b_2 & c_2 & d_2 \\ a_3 & b_3 & c_3 & d_3 \\ a_4 & b_4 & c_4 & d_4 \end{vmatrix},$$

the determinant of the coefficients of the unknowns, and from it the determinants

$$N_x = \begin{vmatrix} k_1 & b_1 & c_1 & d_1 \\ k_2 & b_2 & c_2 & d_2 \\ k_3 & b_3 & c_3 & d_3 \\ k_4 & b_4 & c_4 & d_4 \end{vmatrix}, \quad N_y = \begin{vmatrix} a_1 & k_1 & c_1 & d_1 \\ a_2 & k_2 & c_2 & d_2 \\ a_3 & k_3 & c_3 & d_3 \\ a_4 & k_4 & c_4 & d_4 \end{vmatrix},$$

$$N_z = \begin{vmatrix} a_1 & b_1 & k_1 & d_1 \\ a_2 & b_2 & k_2 & d_2 \\ a_3 & b_3 & k_3 & d_3 \\ a_4 & b_4 & k_4 & d_4 \end{vmatrix}, \quad N_w = \begin{vmatrix} a_1 & b_1 & c_1 & k_1 \\ a_2 & b_2 & c_2 & k_2 \\ a_3 & b_3 & c_3 & k_3 \\ a_4 & b_4 & c_4 & k_4 \end{vmatrix}$$

by replacing the column of coefficients of the indicated unknown by the column of constants.

Cramer's rule states that:

(a) If $D \neq 0$, the system (5-3) has the unique solution

$$x = N_x/D, \qquad y = N_y/D, \qquad z = N_z/D, \qquad w = N_w/D$$

(b) If $D = 0$ and at least one of N_x, N_y, N_z, $N_w \neq 0$, the system has no solution. For, if $D = 0$ and $N_x \neq 0$, then $x \cdot D = N_x$ leads to a contradiction. Such systems are called *inconsistent*.

(c) If $D = 0$ and $N_x = N_y = N_z = N_w = 0$, the system may or may not have a solution. A system having an infinite number of solutions is called *dependent*.

For systems of three or four equations, the simplest procedure is to evaluate D.

Solved Problem 5-5. Solve when possible

(a)
$$2x - 3y + z = 0 \quad (1)$$
$$x + 5y - 3z = 3 \quad (2)$$
$$5x + 12y - 8z = 9 \quad (3)$$

(b)
$$6x - 2y + z = 1 \quad (1)$$
$$x - 4y + 2z = 0 \quad (2)$$
$$4x + 6y - 3z = 0 \quad (3)$$

Solution

(a) Here, $D = 0$; we shall eliminate the variable x.

$$(1) - 2(2): \quad -13y + 7z = -6$$
$$(3) - 5(2): \quad -13y + 7z = -6$$

Then $y = \dfrac{7z + 6}{13}$ and, from (2), $x = 3 - 5y + 3z = \dfrac{4z + 9}{13}$. The

solutions may be written as $x = \dfrac{4a + 9}{13}$, $y = \dfrac{7a + 6}{13}$, $z = a$, where a is arbitrary.

(b) Here $D = 0$; we shall eliminate z.

$$(2) - 2(1): \quad -11x = -2$$
$$(3) + 3(1): \quad 22x = 3$$

The system is inconsistent.

Systems of m Linear Equations in $n > m$ Unknowns

Ordinarily, if there are fewer equations than unknowns, the system will have an infinite number of solutions.

To solve a consistent system of m equations, solve for m of the unknowns (in certain cases, for $p < m$ of the unknowns) in terms of the others.

Systems of n Equations in $m < n$ Unknowns

Ordinarily, if there are more equations than unknowns, the system is inconsistent. However, if $p \leq m$ of the equations have a solution and if this solution satisfies each of the remaining equations, the system is consistent.

A *homogeneous* equation is one in which all terms are of the same degree; otherwise, the equation is called *nonhomogeneous*. For example, the linear equation

$$2x + 3y - 4z = 5$$

is nonhomogeneous, while

$$2x + 3y - 4z = 0$$

is homogeneous.

Every system of homogeneous linear equations

$$a_1 x + b_1 y + c_1 z + \cdots = 0$$
$$a_2 x + b_2 y + c_2 z + \cdots = 0$$

$$\cdot$$

$$\cdot$$

$$\cdot$$

$$a_n x + b_n y + c_n z + \cdots = 0$$

always has the *trivial solution* $x = 0, y = 0, z = 0, \ldots$.

A system of n homogeneous linear equations in n unknowns has *only* the trivial solution if D, the determinant of the coefficients, is not equal to zero. If $D = 0$, the system has nontrivial solutions as well.

Chapter 6
TRIGONOMETRY

I<small>N</small> T<small>HIS</small> C<small>HAPTER</small>:

✔ *Trigonometric Functions of a General Angle*
✔ *Trigonometric Functions of an Acute Angle*
✔ *Reduction to Functions of Positive Acute Angles*
✔ *Variation and Graphs of the Trigonometric Functions*

Trigonometric Functions of a General Angle

Angles in Standard Position

With respect to a rectangular coordinate system, an angle is said to be in *standard position* when its vertex is at the origin and its initial side coincides with the positive *x* axis.

An angle is said to be a *first quadrant angle* or to be *in the first quadrant* if, when in standard position, its terminal side falls in that

107

quadrant. Similar definitions hold for the other quadrants. For example, the angles 30°, 59°, and −330° are first quadrant angles; 119° is a second quadrant angle; −119° is a third quadrant angle; −10° and 710° are fourth quadrant angles. See Figs. 6-1 and 6-2.

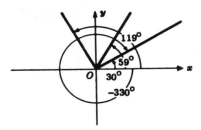

Figure 6-1

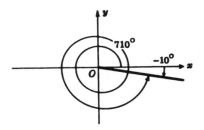

Figure 6-2

Two angles which, when placed in standard position, have coincident terminal sides are called *coterminal angles*. For example, 30° and −330°, −10° and 710° are pairs of coterminal angles. There are an unlimited number of angles coterminal with a given angle.

The angles 0°, 90°, 180°, 270°, and all angles coterminal with them are called *quadrantal angles*.

Trigonometric Functions of a General Angle

Let θ be an angle (not quadrantal) in standard position and let $P(x, y)$ be any point, distinct from the origin, on the terminal side of the angle. The six trigonometric functions of θ are defined, in terms of the abscissa, ordinate and distance of P, as follows:

$$\text{sine } \theta = \sin\theta = \frac{\text{ordinate}}{\text{distance}} = \frac{y}{r}$$

$$\text{cosine } \theta = \cos\theta = \frac{\text{abscissa}}{\text{distance}} = \frac{x}{r}$$

$$\text{tangent } \theta = \tan\theta = \frac{\text{ordinate}}{\text{abscissa}} = \frac{y}{x}$$

$$\text{cosecant } \theta = \csc\theta = \frac{\text{distance}}{\text{ordinant}} = \frac{r}{y}$$

$$\text{secant } \theta = \sec\theta = \frac{\text{distance}}{\text{abscissa}} = \frac{r}{x}$$

$$\text{cotangent } \theta = \cot\theta = \frac{\text{abscissa}}{\text{ordinate}} = \frac{x}{y}$$

Note that $r = \sqrt{x^2 + y^2}$ (see Fig. 6-3).

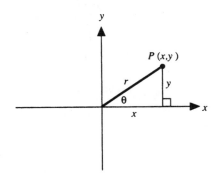

Figure 6-3

As an immediate consequence of these definitions, we have the so-called *reciprocal relations*:

$$\sin\theta = \frac{1}{\csc\theta} \quad \cos\theta = \frac{1}{\sec\theta} \quad \tan\theta = \frac{1}{\cot\theta}$$

$$\csc\theta = \frac{1}{\sin\theta} \quad \sec\theta = \frac{1}{\cos\theta} \quad \cot\theta = \frac{1}{\tan\theta}$$

It is evident from Fig. 6-4(a) – (d) that the values of the trigono-metric functions of θ changes as θ changes. The values of the functions of a given angle θ are, however, independent of the choice of the point P on its terminal side.

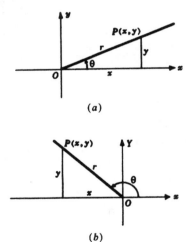

(a)

(b)

Figure 6-4

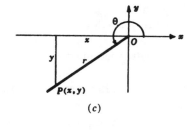

(c)

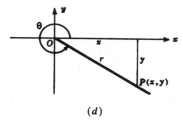

(d)

Figure 6-4

Algebraic Signs of the Functions

Since r is always positive, the signs of the functions in the various quadrants depend upon the signs of x and y. To determine these signs, one may visualize the angle in standard position or use some device as shown in Fig. 6-5 in which the functions having positive signs are listed.

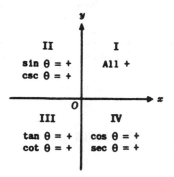

Figure 6-5

When an angle is given, its trigonometric functions are uniquely determined. When, however, the value of one function of an angle is given, the angle is not uniquely determined. For example, if $\sin\theta = \dfrac{1}{2},$ then $\theta = 30^o$, 150°, 390°, 510°,.... . In general, two possible positions of the terminal side are found—for example, the terminal sides of 30° and 150° in figure 6-4(a) and (b). The exceptions to this rule occur when the angle is quadrantal.

Trigonometric Functions of Quadrantal Angles

For a quadrantal angle, the terminal side coincides with one of the axes. A point P, distinct from the origin, on the terminal side has either $x = 0$, $y \neq 0$, or $x \neq 0$, $y = 0$. In either case, two of the six functions will not be defined. For example, the terminal side of the angle 0° coincides with the positive x axis and the ordinate of P is 0. Since the ordinate occurs in the denominator of the ratio defining the cotangent and cosecant, these functions are not defined. The trigonometric functions of the quadrantal angles are given in Table 6-1.

angle θ	$\sin\theta$	$\cos\theta$	$\tan\theta$	$\cot\theta$	$\sec\theta$	$\csc\theta$
0°	0	1	0		1	undefined
90°	1	0	undefined	0	undefined	1
180°	0	-1	0		-1	undefined
270°	-1	0	undefined	0	undefined	-1

Table 6-1

Trigonometric Functions of an Acute Angle

In dealing with any right triangle, it will be convenient (see Fig. 6-6) to denote the vertices as A, B, C such that C is the vertex of the right triangle; to denote the angles of the triangle as A, B, C such that $m\angle C = 90^{\circ}$; and the sides opposite the angles as a, b, c, respectively. With respect to angle A, a will be called the *opposite side* and b the *adjacent side*. With respect to angle B, a will be called the *adjacent side* and b the *opposite side*. Side c will always be called the *hypotenuse*.

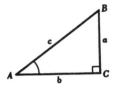

Figure 6-6

If now the right triangle is placed in a coordinate system (Fig. 6-7) so that angle A is in standard position, point B on the terminal side of angle A has coordinates (b, a) and distance $c = \sqrt{a^2 + b^2}$. Then the trigonometric functions of angle A may be defined in terms of the sides of the right triangle as follows:

$$\sin A = \frac{a}{c} = \frac{\text{length of opposite side}}{\text{length of hypotenuse}}$$

$$\cos A = \frac{b}{c} = \frac{\text{length of adjacent side}}{\text{length of hypotenuse}}$$

$$\tan A = \frac{a}{b} = \frac{\text{length of opposite side}}{\text{length of adjacent side}}$$

$$\csc A = \frac{c}{a} = \frac{\text{length of hypotenuse}}{\text{length of opposite side}}$$

$$\sec A = \frac{c}{b} = \frac{\text{length of hypotenuse}}{\text{length of adjacent side}}$$

$$\cot A = \frac{b}{a} = \frac{\text{length of adjacent side}}{\text{length of opposite side}}$$

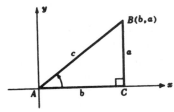

Figure 6-7

Trigonometric Functions of Complementary Angles

The acute angles A and B of the right triangle ABC are complementary, that is, $m\angle A + m\angle B = 90^\circ$. From Fig. 6-6, we have

$$\sin B = \frac{b}{c} = \cos A \qquad \csc B = \frac{c}{b} = \sec A$$

$$\cos B = \frac{a}{c} = \sin A \qquad \sec B = \frac{c}{a} = \csc A$$

$$\tan B = \frac{b}{a} = \cot A \qquad \cot B = \frac{a}{b} = \tan A$$

These relations associate their functions in pairs–sine and cosine, tangent and cotangent, secant and cosecan–teach function of a pair being called the *cofunction* of the other. Thus, any function of an acute angle is equal to the corresponding cofunction of the complementary angle.

Trigonometric Functions of 30°, 45°, 60°

Angle θ	$\sin \theta$	$\cos \theta$	$\tan \theta$	$\cot \theta$	$\sec \theta$	$\csc \theta$
30°	$\frac{1}{2}$	$\frac{1}{2}\sqrt{3}$	$\frac{1}{3}\sqrt{3}$	$\sqrt{3}$	$\frac{2}{3}\sqrt{3}$	2
45°	$\frac{1}{2}\sqrt{2}$	$\frac{1}{2}\sqrt{2}$	1	1	$\sqrt{2}$	$\sqrt{2}$
60°	$\frac{1}{2}\sqrt{3}$	$\frac{1}{2}$	$\sqrt{3}$	$\frac{1}{3}\sqrt{3}$	2	$\frac{2}{3}\sqrt{3}$

Table 6-2

Solved Problem 6-1. Find the values of the trigonometric functions of 30° and 60°.

Solution. In any equilateral triangle ABD (see Fig. 6-8), each angle is 60°. The bisector of any angle, as B, is the perpendicular bisector of the opposite side. Let the sides of the equilateral triangle be of length 2 units. Then, in the right triangle ABC, $AB = 2$, $AC = 1$, and

$$BC = \sqrt{2^2 - 1^2} = \sqrt{3}.$$

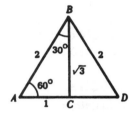

Figure 6-8

$$\sin 30° = \frac{1}{2} = \cos 60° \qquad\qquad \csc 30° = 2 = \sec 60°$$

$$\cos 30° = \frac{\sqrt{3}}{2} = \sin 60° \qquad\qquad \sec 30° = \frac{2}{\sqrt{3}} = \frac{2\sqrt{3}}{3} = \csc 60°$$

$$\tan 30° = \frac{1}{\sqrt{3}} = \frac{\sqrt{3}}{3} = \cot 60° \qquad \cot 30° = \sqrt{3} = \tan 60°$$

Angle θ	sin θ	cos θ	tan θ	cot θ	sec θ	csc θ
15°	0.26	0.97	0.27	3.7	1.0	3.9
20°	0.34	0.94	0.36	2.7	1.1	2.9
30°	0.50	0.87	0.58	1.7	1.2	2.0
40°	0.64	0.77	0.84	1.2	1.3	1.6
45°	0.71	0.71	1.0	1.0	1.4	1.4
50°	0.77	0.64	1.2	0.84	1.6	1.3
60°	0.87	0.50	1.7	0.58	2.0	1.2
70°	0.94	0.34	2.7	0.36	2.9	1.1
75°	0.97	0.26	3.7	0.27	3.9	1.0

Table 6-3

Solved Problem 6-2. When the sun is 20° above the horizon, how long is the shadow cast by a building 150 ft high.

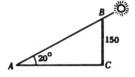

Figure 6-9

Solution. In Fig. 6-9, $A = 20°$ and $CB = 150$. Then $\cot A = AC/CB$ and $AC = CB \cot A = 150 \cot 20° = 150(2.7) = 405$ ft.

Reduction to Functions of Positive Acute Angles

Coterminal Angles

Let θ be any angle; then

$$\sin\left(\theta + n360°\right) = \sin\theta \qquad \csc\left(\theta + n360°\right) = \csc\theta$$

$$\cos\left(\theta + n360°\right) = \cos\theta \qquad \sec\left(\theta + n360°\right) = \sec\theta$$

$$\tan\left(\theta + n360°\right) = \tan\theta \qquad \cot\left(\theta + n360°\right) = \cot\theta$$

where n is any positive or negative integer or zero.

Example 6-1

$$\sin 400° = \sin\left(40° + 360°\right) = \sin 40°$$

$$\cos 850° = \cos\left(130° + 2 \cdot 360°\right) = \cos 130°$$

$$\tan\left(-1000°\right) = \tan\left(80° - 3 \cdot 360°\right) = \tan 80°$$

Functions of a Negative Angle

Let θ be any angle; then

$$\sin(-\theta) = -\sin\theta \qquad \csc(-\theta) = -\csc\theta$$
$$\cos(-\theta) = \cos\theta \qquad \sec(-\theta) = \sec\theta$$
$$\tan(-\theta) = -\tan\theta \qquad \cot(-\theta) = -\cot\theta$$

Example 6-2

$$\sin(-50°) = -\sin 50°$$
$$\cos(-30°) = \cos 30°$$
$$\tan(-200°) = -\tan(200°)$$

Reduction Formulas

Let θ be any angle; then

$$\sin(90° - \theta) = \cos\theta \qquad \sin(90° + \theta) = \cos\theta$$
$$\cos(90° - \theta) = \sin\theta \qquad \cos(90° + \theta) = -\sin\theta$$
$$\tan(90° - \theta) = \cot\theta \qquad \tan(90° + \theta) = -\cot\theta$$
$$\csc(90° - \theta) = \sec\theta \qquad \csc(90° + \theta) = \sec\theta$$
$$\sec(90° - \theta) = \csc\theta \qquad \sec(90° + \theta) = -\csc\theta$$
$$\cot(90° - \theta) = \tan\theta \qquad \cot(90° + \theta) = -\tan\theta$$

$$\sin\left(180^\circ - \theta\right) = \sin\theta \qquad \sin\left(180^\circ + \theta\right) = -\sin\theta$$

$$\cos\left(180^\circ - \theta\right) = -\cos\theta \qquad \cos\left(180^\circ + \theta\right) = -\cos\theta$$

$$\tan\left(180^\circ - \theta\right) = -\tan\theta \qquad \tan\left(180^\circ + \theta\right) = \tan\theta$$

$$\csc\left(180^\circ - \theta\right) = \csc\theta \qquad \csc\left(180^\circ + \theta\right) = -\csc\theta$$

$$\sec\left(180^\circ - \theta\right) = -\sec\theta \qquad \sec\left(180^\circ + \theta\right) = -\sec\theta$$

$$\cot\left(180^\circ - \theta\right) = -\cot\theta \qquad \cot\left(180^\circ + \theta\right) = \cot\theta$$

General Reduction Formula

Any trigonometric function of $\left(n \cdot 90^\circ \pm \theta\right)$, where θ is any angle, is *numerically* equal

(a) to the same function of θ if n is an even integer

(b) to the corresponding cofunction of θ if n is an odd integer

The algebraic sign in each case is the same as the sign of the given function for that quadrant in which $\left(n \cdot 90^\circ \pm \theta\right)$ lies when θ is a positive acute angle.

Example 6-3.

(a) $\sin\left(180^\circ - \theta\right) = \sin\left(2 \cdot 90^\circ - \theta\right) = \sin\theta$ since 180° is an even multiple of 90° and, when q is positive acute, the terminal side of $180^\circ - \theta$ lies in quadrant II.

(b) $\cos\left(180^\circ + \theta\right) = \cos\left(2 \cdot 90^\circ + \theta\right) = -\cos\theta$ since 180° is an even multiple of 90° and, when θ is positive acute, the terminal side of $180^\circ + \theta$ lies in quadrant III.

Variations and Graphs of the Trigonometric Functions

Line Representations of the Trigonometric Functions

Let θ be any given angle in standard position. (See Figs. 6-10 through 6-13 for θ in each of the quadrants.)

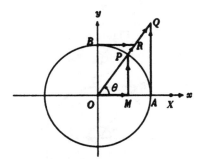

Figure 6-10

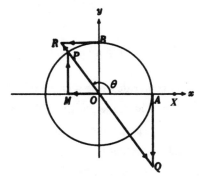

Figure 6-11

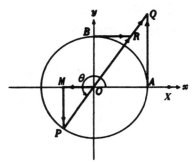

Figure 6-12

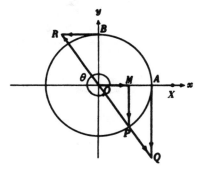

Figure 6-13

With the vertex O as center describe a circle of radius one unit cutting the initial side $\overline{OX}$ of θ at A, the positive y axis at B, and the terminal side of θ at P. Draw $\overline{MP}$ perpendicular to $\overline{OX}$, draw also the tangents to the circle at A and B meeting the terminal side of θ or its extension through O in the points Q and R, respectively.

In each of the figures, the right triangles OMP, OAQ, and OBR are similar, and

$$\sin\theta = \frac{MP}{OP} = MP \qquad \csc\theta = \frac{OP}{MP} = \frac{OR}{OB} = OR$$

$$\cos\theta = \frac{OM}{OP} = OM \qquad \sec\theta = \frac{OP}{OM} = \frac{OQ}{OA} = OQ$$

$$\tan\theta = \frac{MP}{OM} = \frac{AQ}{OA} = AQ \qquad \cot\theta = \frac{OM}{MP} = \frac{BR}{OB} = BR$$

Then, $\overline{MP}, \overline{OM}, \overline{AQ}$, etc., are directed line segments, the magnitude of a function being given by the length of the corresponding segment and the sign being given by the indicated direction. The directed segments $\overline{OQ}$ and $\overline{OR}$ are to be considered positive when measured on the terminal side of the angle and negative when measured on the terminal side extended.

Variations of the Trigonometric Functions

Let P move counterclockwise about the unit circle, starting at A, so that $m\angle\theta = m\angle XOP$ varies continuously from $0°$ to $360°$. Using Figs. 6-10 to 6-13, Table 6-4 is derived.

As θ increases from	0° to 90°	90° to 180°	180° to 270°	270° to 360°
sin θ	I from 0 to 1	D from 1 to 0	D from 0 to -1	I from -1 to 0
cos θ	D from 1 to 0	D from 0 to -1	I from -1 to 0	I from 0 to 1
tan θ	I from 0 without limit (0 to $+\infty$)	I from large negative values to 0 ($-\infty$ to 0)	I from 0 without limit (0 to $+\infty$)	I from large negative values to 0 ($-\infty$ to 0)
cot θ	D from large positive values to 0 ($+\infty$ to 0)	D from 0 without limit (0 to $-\infty$)	D from large positive values to 0 ($+\infty$ to 0)	D from 0 without limit (0 to $-\infty$)
sec θ	I from 1 without limit (1 to $+\infty$)	I from large negative values to -1 ($-\infty$ to -1)	D from -1 without limit (-1 to $-\infty$)	D from large positive values to 1 ($+\infty$ to 1)
csc θ	D from large positive values to 1 ($+\infty$ to 1)	I from 1 without limit (1 to $+\infty$)	I from large negative values to -1 ($-\infty$ to -1)	D from -1 without limit (-1 to $-\infty$)

I = increases; D = decreases.

Table 6-4

Graphs of the Trigonometric Functions

In Table 6-5, values of the angle x are given in radians.

x	$y = \sin x$	$y = \cos x$	$y = \tan x$	$y = \cot x$	$y = \sec x$	$y = \csc x$
0	0	1.00	0	$\pm\infty$	1.00	$\pm\infty$
$\pi/6$	0.50	0.87	0.58	1.73	1.15	2.00
$\pi/4$	0.71	0.71	1.00	1.00	1.41	1.41
$\pi/3$	0.87	0.50	1.73	0.58	2.00	1.15
$\pi/2$	1.00	0	$\pm\infty$	0	$\pm\infty$	1.00
$2\pi/3$	0.87	−0.50	−1.73	−0.58	−2.00	1.15
$3\pi/4$	0.71	−0.71	−1.00	−1.00	−1.41	1.41
$5\pi/6$	0.50	−0.87	−0.58	−1.73	−1.15	2.00
π	0	−1.00	0	$+\infty$	−1.00	$+\infty$
$7\pi/6$	−0.50	−0.87	0.58	1.73	−1.15	−2.00
$5\pi/4$	−0.71	−0.71	1.00	1.00	−1.41	−1.41
$4\pi/3$	−0.87	−0.50	1.73	0.58	−2.00	−1.15
$3\pi/2$	−1.00	0	$\pm\infty$	0	$\pm\infty$	−1.00
$5\pi/3$	−0.87	0.50	−1.73	−0.58	2.00	−1.15
$7\pi/4$	−0.71	0.71	−1.00	−1.00	1.41	−1.41
$11\pi/6$	−0.50	0.87	−0.58	−1.73	1.15	−2.00
2π	0	1.00	0	$\pm\infty$	1.00	$\pm\infty$

Table 6-5

Note 1. Since $\sin\left(\dfrac{1}{2}\pi + x\right) = \cos x,$ the graph of $y = \cos x$ may be obtained most easily by shifting the graph of $y = \sin x$ a distance $\dfrac{1}{2}\pi$ to the left. See Fig. 6-14.

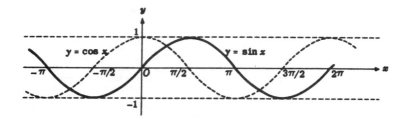

Figure 6-14

Note 2. Since $\csc\left(\dfrac{1}{2}\pi + x\right) = \sec x$, the graph of $y = \csc x$ may be obtained by shifting the graph of $y = \sec x$ a distance $\dfrac{1}{2}\pi$ to the right. Notice, too, the relationship between the graphs for $\tan x$ and $\cot x$. See Fig. 6-15 through 6-18.

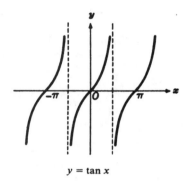

$y = \tan x$

Figure 6-15

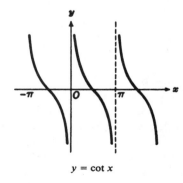

$y = \cot x$

Figure 6-16

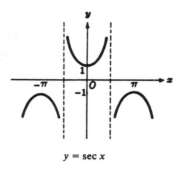

$y = \sec x$

Figure 6-17

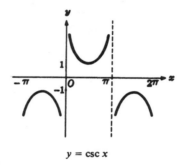

$y = \csc x$

Figure 6-18

Periodic Functions

sAny function of a variable x, $f(x)$, which repeats itself in definite cycles, is called *periodic*. The smallest range of values of x which corresponds to a complete cycle of values of the function is called the period of the function. It is evident from the graphs of the trigonometric functions that the sine, cosine, secant, and cosecant are of period 2π while the tangent and cotangent are of period π.

The General Sine Curve

The *amplitude* (*height*) and *period* (*wavelength*) of $y = \sin x$ are, respectively, 1 and 2π. For a given value of x, the value of $y = a \sin x$, $a > 0$, is a times the value of $y = \sin x$. Thus, the amplitude of $y = a \sin x$ is a and the period is 2π. Since, when $bx = 2\pi$, $x = 2\pi/b$, the amplitude of $y = \sin bx$, $b > 0$, is 1 and the period is $2\pi/b$.

The general sine curve (sinusoid) of equation

$$y = a \sin bx, \qquad\qquad a > 0, \quad b > 0,$$

has amplitude a and period $2\pi/b$. Thus, the graph of $y = 3 \sin 2x$ has amplitude 3 and period $2\pi/2 = \pi$. Fig. 6-19 exhibits the graphs of $y = \sin x$ and $y = 3 \sin 2x$ on the same axes.

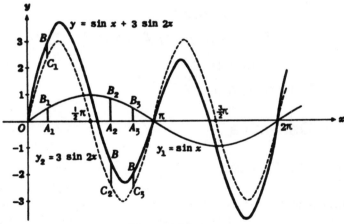

Figure 6-19

Chapter 7
INTRODUCTION TO CALCULUS

IN THIS CHAPTER:

✔ *The Derivative*
✔ *Differentiation of Algebraic Expressions*
✔ *Integration*

The Derivative

Limit of a Function

A given function $f(x)$ is said to have a *limit M* as x approaches c [in symbols, $\lim_{x \to c} f(x) = M$] if $f(x)$ can be made as close to M for all values $x \neq c$ but sufficiently near to c, by having x get sufficiently close to c (approaching both from the left and right).

Example 7-1. Consider $f(x) = \dfrac{x^2 - x - 6}{x - 3}$ for x near 3. When $x \neq 3$,

$f(x) = \dfrac{x^2 - x - 6}{x - 3} = \dfrac{(x-3)(x+2)}{(x-3)} = (x+2)$. Thus, for x near 3, $x + 2$

is near 5 and

$$\lim_{x \to 3} \frac{x^2 - x - 6}{x - 3} = 5$$

Theorems on Limits

If $\lim_{x \to c} f(x) = M$ and $\lim_{x \to c} g(x) = N$, then

I. $\lim_{x \to c} \left[f(x) \pm g(x) \right] = \lim_{x \to c} f(x) \pm \lim_{x \to c} g(x) = M \pm N$

II. $\lim_{x \to c} \left[kf(x) \right] = k \lim_{x \to c} f(x) = kM$, where k is a constant.

III. $\lim_{x \to c} \left[f(x) \cdot g(x) \right] = \lim_{x \to c} f(x) \cdot \lim_{x \to c} g(x) = MN$

IV. $\lim_{x \to c} \dfrac{f(x)}{g(x)} = \dfrac{\lim\limits_{x \to c} f(x)}{\lim\limits_{x \to c} g(x)} = \dfrac{M}{N}$, provided $N \neq 0$.

Continuous Functions

A function $f(x)$ is called *continuous* at $x = c$, provided

(1) $f(c)$ is defined,

(2) $\lim\limits_{x \to c} f(x)$ exists,

(3) $\lim\limits_{x \to c} f(x) = f(c)$.

Example 7-2. The function $f(x) = \dfrac{x^2 - x - 6}{x - 3}$ is not continuous at $x = 3$ since $f(3)$ is not defined.

A function $f(x)$ is said to be *continuous* on the interval (a, b) if it is continuous for every value of x of the interval. A polynomial in x is continuous since it is continuous for all values of x. A rational function in x, $f(x) = P(x) / Q(x)$, where $P(x)$ and $Q(x)$ are polynomials, is continuous for all values of x except those for which $Q(x) = 0$. Thus, $f(x) = \dfrac{x^2 + x + 1}{(x-1)(x^2 + 2)}$ is continuous for all values of x, except $x = 1$.

The Derivative

The *derivative* of $y = f(x)$ at $x = x_o$ is

$$\lim_{\Delta x \to 0} \frac{\Delta y}{\Delta x} = \lim_{\Delta x \to 0} \frac{f(x_o + \Delta x) - f(x_o)}{\Delta x}$$

provided the limit exists.

In finding derivatives, we shall use the following five-step rule:

(1) Write $y_o = f(x_o)$.

(2) Write $y_o + \Delta y = f(x_o + \Delta x)$.

(3) Obtain $\Delta y = f\left(x_o + \Delta x\right) - f\left(x_o\right)$.

(4) Obtain $\dfrac{\Delta y}{\Delta x}$.

(5) Evaluate $\lim\limits_{\Delta x \to 0} \dfrac{\Delta y}{\Delta x}$.

Solved Problem 7-1. Find the derivative of $y = f\left(x\right) = 2x^2 - 3x + 5$ at $x = x_o$.

Solution

(1) $y_o = f\left(x_o\right) = 2x_o^2 - 3x_o + 5$

(2) $\begin{aligned} y_o + \Delta y = f\left(x_o + \Delta x\right) &= 2\left(x_o + \Delta x\right)^2 - 3\left(x_o + \Delta x\right) + 5 \\ &= 2x_o^2 + 4x_o \cdot \Delta x + 2\left(\Delta x\right)^2 - 3x_o - 3 \cdot \Delta x + 5 \end{aligned}$

(3) $\Delta y = f\left(x_o + \Delta x\right) - f\left(x_o\right) = 4x_o \cdot \Delta x - 3 \cdot \Delta x + 2\left(\Delta x\right)^2$

(4) $\dfrac{\Delta y}{\Delta x} = 4x_o - 3 + 2 \cdot \Delta x$

(5) $\lim\limits_{\Delta x \to 0} \dfrac{\Delta y}{\Delta x} = \lim\limits_{\Delta x \to 0}\left(4x_o - 3 + 2 \cdot \Delta x\right) = 4x_o - 3$

If, in the problem above, the subscript 0 is deleted, the five-step rule yields a function of x (here, $4x - 3$) called the derivative with respect to x of the given function. The derivative with respect to x of the function

$y = f(x)$ is denoted by one of the symbols y', $\dfrac{dy}{dx}$, $f'(x)$, or $D_x y$.

Higher-Order Derivatives

The process of finding the derivative of a given function is called *differentiation*. By differentiation, we obtain from a given function

$y = f(x)$ another function $y' = f'(x)$ which will now be called the

first derivative of y or of $f(x)$ with respect to x. If, in turn, $y' = f'(x)$

is differentiated with respect to x, another function $y'' = f''(x)$, called

the *second derivative* of y or of $f(x)$ is obtained. Similarly, a third derivative may be found, and so on.

Solved Problem 7-2. Given the function $y = f(x) = x^4 - 3x^2 + 8x + 6$, find the: (a) first derivative, (b) second derivative, and (c) third derivative.

Solution

(a) $y' = f'(x) = 4x^3 - 6x + 8$

(b) $y'' = f''(x) = 12x^2 - 6$

(c) $y''' = f'''(x) = 24x$

Differentiation of Algebraic Expressions

Differentiation Formulas

I.   If $y = f(x) = kx^n$, where k and n are constants, then

$y' = f'(x) = knx^{n-1}$.

II. If $y = k \cdot u^n$, where k and n are constants and u is a function

of x, then $y' = knu^{n-1} \cdot u'$, provided u' exists. This is the *chain rule*.

Solved Problem 7-3. Find y', given $y = 8x^{5/4}$.

Solution. Here $k = 8$, $n = \dfrac{5}{4}$. Then, $y' = knx^{n-1} = 8 \cdot \dfrac{5}{4} x^{(5/4)-1} = 10x^{1/4}$

III. If $y = f(x) \cdot g(x)$, then $y' = f(x) \cdot g'(x) + g(x) \cdot f'(x)$, provided $f'(x)$ and $g'(x)$ exist.

Solved Problem 7-4. Find y' when $y = \left(x^3 + 3x^2 + 1\right)\left(x^2 + 2\right)$.

Solution. Take $f(x) = x^3 + 3x^2 + 1$ and $g(x) = x^2 + 2$. Then, $f'(x) = 3x^2 + 6x$, $g'(x) = 2x$, and

$$y' = f(x) \cdot g'(x) + g(x) \cdot f'(x)$$
$$= \left(x^3 + 3x^2 + 1\right)\left(2x\right) + \left(x^2 + 2\right)\left(3x^2 + 6x\right)$$
$$= 5x^4 + 12x^3 + 6x^2 + 14$$

IV. If $y = \dfrac{f(x)}{g(x)}$, then $y' = \dfrac{g(x) \cdot f'(x) - f(x) \cdot g'(x)}{\left[g(x)\right]^2}$, when

$f'(x)$ and $g'(x)$ exist and $g(x) \neq 0$.

Solved Problem 7-5. Find y', given $y = \dfrac{x+1}{x^2+1}$.

Solution. Take $f(x) = x+1$ and $g(x) = x^2 + 1$. Then

$$y' = \frac{g(x) \cdot f'(x) - f(x) \cdot g'(x)}{\left[g(x)\right]^2} = \frac{\left(x^2+1\right)\left(1\right) - \left(x+1\right)\left(2x\right)}{\left(x^2+1\right)^2} = \frac{1 - 2x - x^2}{\left(x^2+1\right)^2}$$

Integration

If $F(x)$ is a function whose derivative $F'(x) = f(x)$, then $F(x)$ is

called an *integral* of $f(x)$. For example, $F(x) = x^3$ is an integral of

$f(x) = 3x^2$ since $F'(x) = 3x^2 = f(x)$. Also, $G(x) = x^3 + 5$ and

$H(x) = x^3 - 6$ are integrals of $f(x) = 3x^2$. Why?

If $F(x)$ and $G(x)$ are two distinct integrals of $f(x)$, then $F(x) = G(x) + C$, where C is a constant.

The Indefinite Integral

The indefinite integral or antiderivative of $f(x)$, denoted by $\int f(x)\,dx$, is the most general integral of $f(x)$, that is, $\int f(x)\,dx = F(x) + C$ where $F(x)$ is any function such that $F'(x) = f(x)$ and C is an arbitrary constant. Thus, the indefinite integral of $f(x) = 3x^2$ is

$$\int 3x^2\,dx = x^3 + C$$

We shall use the following antidifferentiation formulas:

I. $\int x^n\,dx = \dfrac{x^{n+1}}{n+1} + C$, where $n \neq -1$

II. $\int cf(x)\,dx = c\int f(x)\,dx$, where c is a constant

III. $\int [f(x) + g(x)]\,dx = \int f(x)\,dx + \int g(x)\,dx$

Example 7-3

(a) $\int x^5\,dx = \dfrac{x^{5+1}}{5+1} + C = \dfrac{x^6}{6} + C$

(b) $\int 4x^3\,dx = 4\int x^3\,dx = 4 \cdot \dfrac{x^4}{4} + C = x^4 + C$

(c) $\int 3x\,dx = 3\int x\,dx = 3 \cdot \dfrac{x^2}{2} + C = \dfrac{3}{2}x^2 + C$

(d) $\quad \int \dfrac{dx}{x^3} = \int x^{-3}\ dx = \dfrac{x^{-2}}{-2} + C = -\dfrac{1}{2x^2} + C$

Area by Summation

Consider the area A bounded by the curve $y = f(x) \geq 0$, the x axis, and the ordinates $x = a$ and $x = b$, where $b > a$.

Let the interval $a \leq x \leq b$ be divided into n equal parts each of length Δx. At each point of subdivision, construct the ordinate thus dividing the area into n strips, as in Fig. 7-1.

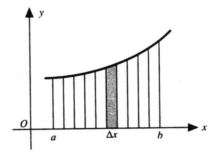

Figure 7-1

Since the area of the strips are unknown, we propose to approximate each strip by a rectangle whose area can be found. In Fig. 7-2, a representative strip and its approximating rectangle are shown.

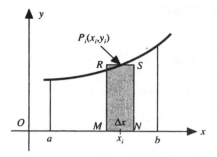

Figure 7-2

Suppose the representative strip is the ith strip counting from the left, and let $x = x_i$ be the coordinate of the midpoint of its base. Denote by $y_i = f(x_i)$ the ordinate of the point P_i (on the curve) whose abscissa is x_i. Through P_i pass a line parallel to the x axis and complete the rectangle $MRSN$. This rectangle of area $y_i \Delta x$ is the approximating rectangle of the ith strip. When each strip is treated similarly, it seems reasonable to take

$$y_1 \Delta x + y_2 \Delta x + y_3 \Delta x + \cdots + y_n \Delta x = \sum_{i=1}^{n} y_i \Delta x$$

as an approximation of the area sought.

Now suppose that the number of strips (with approximating rectangles) is indefinitely increased so that $\Delta x \to 0$. It is evident from the figure that by increasing the number of approximating rectangles, the sum of their areas more nearly approximates the area sought, that is,

$$A = \lim_{n \to \infty} \sum_{i=1}^{n} y_i \Delta x$$

The Definite Integral

If we define $\int_a^b f(x)\,dx$ [read, the *definite integral* of $f(x)$ between $x = a$ and $x = b$] as follows

$$\int_a^b f(x)\,dx = F(x)\big|_a^b = F(b) - F(a), \qquad (F'(x) = f(x))$$

then the area bounded by $y = f(x) \geq 0$, the x axis, and the ordinates $x = a$ and $x = b$, $(b > a)$, is given by

$$A = \int_a^b f(x)\,dx$$

Index